Science Mysteries Explained

by Anthony Fordham

A member of Penguin Group (USA) Inc.

ALPHA BOOKS

Published by Penguin Group (USA) Inc.

Penguin Group (USA) Inc., 375 Hudson Street, New York, New York 10014, USA · Penguin Group (Canada), 90 Eglinton Avenue East, Suite 700, Toronto, Ontario M4P 2Y3, Canada (a division of Pearson Penguin Canada Inc.) · Penguin Books Ltd., 80 Strand, London WC2R 0RL, England · Penguin Ireland, 25 St. Stephen's Green, Dublin 2, Ireland (a division of Penguin Books Ltd.) · Penguin Group (Australia), 250 Camberwell Road, Camberwell, Victoria 3124, Australia (a division of Pearson Australia Group Pty. Ltd.) · Penguin Books India Pvt. Ltd., 11 Community Centre, Panchsheel Park, New Delhi—110 017, India · Penguin Group (NZ), 67 Apollo Drive, Rosedale, North Shore, Auckland 1311, New Zealand (a division of Pearson New Zealand Ltd.) · Penguin Books (South Africa) (Pty.) Ltd., 24 Sturdee Avenue, Rosebank, Johannesburg 2196, South Africa · Penguin Books Ltd., Registered Offices: 80 Strand, London WC2R 0RL, England

International Standard Book Number: 978-1-61564-458-2
Library of Congress Catalog Card Number: 2013956274

16 15 14 8 7 6 5 4 3 2 1

Interpretation of the printing code: The rightmost number of the first series of numbers is the year of the book's printing; the rightmost number of the second series of numbers is the number of the book's printing. For example, a printing code of 14-1 shows that the first printing occurred in 2014.

Publisher: *Mike Sanders*
Executive Managing Editor: *Billy Fields*
Senior Acquisitions Editor: *Tom Stevens*
Development Editor: *John Etchison*
Senior Production Editor: *Janette Lynn*
Cover Designer: *Rebecca Batchelor*
Illustrator: *John Frasier*
Indexer: *Angie Bess Martin*
Proofreader: *Gene Redding*

contents

life science 56

cosmology . . . 164

physics 218

introduction

Humans are curious creatures. For thousands of years, we've looked at the stars, the sea, the earth beneath our feet, and the other creatures we share this planet with and thought: What gives? For thousands of years, we told each other fanciful stories about how the Earth was created, and how and why the things in it interact. Then, just a few short centuries ago, we came up with a new idea, a new way of describing the world: science.

Science is how we come to understand everything around us in a way that's consistent and sensible. But science is anything but straightforward. In fact, sometimes it seems to just raise an endless series of questions!

Hopefully, then, this book will give you some of the answers to those questions. We've chosen some of the biggest or most vexing questions in science and answered them in a way that's clear and straightforward.

After each question, we've written a short paragraph that explains the question in a little more detail. Then we give you a short answer, so you can get a quick sense of what you're in for (especially useful in the Cosmology section)! Then the full answer is supported by illustrations that will make everything crystal clear, and you the font of all knowledge among your friends!

Because science encompasses, well, everything, we've split the book into easily digestible chunks, as follows:

Earth Science
Everything about the planet beneath us. What the Earth is made of, how it formed, and what its ultimate fate might be. From earthquakes to ice caps, climate change to volcanoes and why our day is 24 hours long.

Life Science
We live on the only life-bearing world ... as far as we know. So what makes life alive? Where did it come from, and where is it going? How does evolution really work, and why can't we make our food from the Sun like plants? All this and more!

Chemistry
You might dimly remember from schooldays the Periodic Table of Elements, or something about molecules and chemical bonds. Here, it's all explained—how atoms connect to each other and store energy in those connections, and how that simple idea makes the whole world work!

Cosmology
Most of everything is out there, in deep space, being all weird and ridiculously huge and far away. The universe might not be infinite, but it's so huge we might as well *say* it's infinite. If you have questions about stars, moons, planets, and more, this is the place to find answers.

Physics
The laws of nature are powerful, but do you know them as well as you should? Is it even possible to go back in time to take those classes all over again? From surviving lightning strikes to floating in the ocean to being swallowed by a black hole—the answers are all found in physics.

Special Thanks to the Technical Reviewer

Idiot's Guides: Science Mysteries Explained was reviewed by an expert who double-checked the accuracy of what you'll learn here, to help us ensure this book gives you everything you need to know about these mysteries of science. Special thanks are extended to Nicholas Reid.

earth science

What makes our tiny ball of rock and water so special?

The planet beneath us is what makes the world around us. Earth is the only life-bearing planet we know of, but its internal structure and the systems that power it took us centuries of careful study to work out.

Without understanding how our planet works, we wouldn't have a hope of figuring out how life works. From Earth's volcanoes and earthquakes to the peculiarities of its orbit and the mix of minerals in the crust—all these things combine to make the home we love.

What is it that makes Earth so special? Why is this the place we evolved, rather than Mars or Venus or one of the gas giants? Do we owe our lives to Earth's magnetic field? And why is our civilization so dependent on a random mix of chemicals scattered across the surface and the upper crust?

Our world is all we have, right now, and understanding it could mean the difference between thousands of years of prosperity ... or extinction.

How do we know the Earth is 4.54 billion years old?

Figuring out the age of the planet we live on is one of the biggest questions in science. For centuries, different theories have steadily increased the assumed age of the Earth. How can we be sure we're right now?

A:

Radioactive elements like uranium decay into stable elements like lead over a specific amount of time. By measuring the amount of lead in a sample of uranium, scientists can calculate how old the uranium is. But that's just the easy part ....

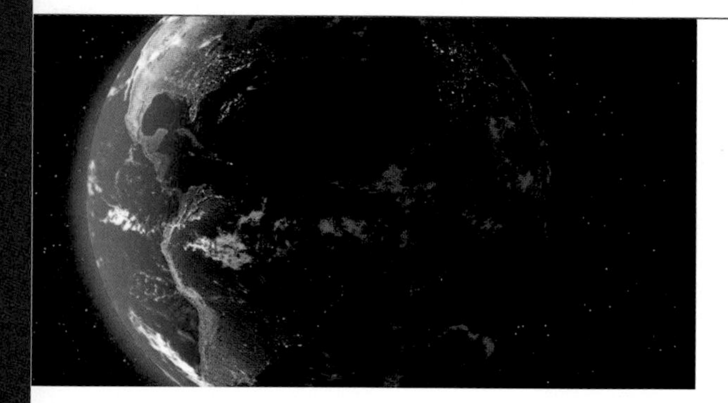

Early on after the scientific revolution in the seventeenth century, scientists already knew enough about layered rocks, fossils, and other clues to make them suspect the Earth was many tens of thousands—perhaps millions—of years old.

But back then, humans did not know the interior of the Earth was liquid, nor did they understand the process of radioactive decay. Without this vital knowledge, their models and ideas of how the Earth formed were hugely flawed.

The first scientific theories of the age of the Earth calculated how long a planet of our size would take to cool and solidify from its initial molten state. By measuring the temperature of rocks and making estimates about the size of the Sun, scientists came up with figures of anywhere from 75,000 to 20 million years old.

But there was a big problem with their theories: because of swirling molten magma inside the Earth and nuclear fusion inside the Sun, the rate of cooling is much slower than you might expect. In other words, the Earth remains much hotter than it would if it had just formed in space and cooled.

By the mid-nineteenth century, scientists had developed a better understanding of what the interior of the Earth was like. They knew about the fluid interior and the constant upwelling of magma. Their estimates of the age of the planet changed to the range of tens of millions to hundreds of millions of years old.

But there were still two vital parts of the puzzle missing: the constant renewal of the surface of the Earth over millions of years through continental drift and the theory of radioactive decay.

Scientists initially assumed that a layer of rock in the Earth had been there since the planet formed. Eventually, they realized the surface changed, and many rocks on the surface had been melted and reformed. Just looking at surface rocks wasn't the way to calculate an accurate age for the Earth.

The final breakthrough came in the late nineteenth century with the discovery of radioactivity. Scientists discovered certain elements—especially uranium—decayed at a constant rate. They knew that if they analyzed, say, a pound of uranium, they could count how much lead had formed inside the lump via radioactive decay.

Imagine putting a scoop of ice cream onto a plate and noting that it takes one hour for the ice cream to completely melt. Now, the next time you see a scoop of ice cream on a plate, you can measure how much of the ice cream has already melted—say, half—and estimate that the ice cream must have been scooped out half an hour ago.

The same principle applies to the "radiometric" dating of rocks. Scientists can figure out a maximum age for the rock and make the assumption the rock cannot be older than the planet itself.

The oldest rocks we've so far measured are 4.54 billion years old. The accuracy of the model is refined further by combining our theories of how the Solar System formed and the characteristics of special meteorites to get an upper limit for the age of our planet.

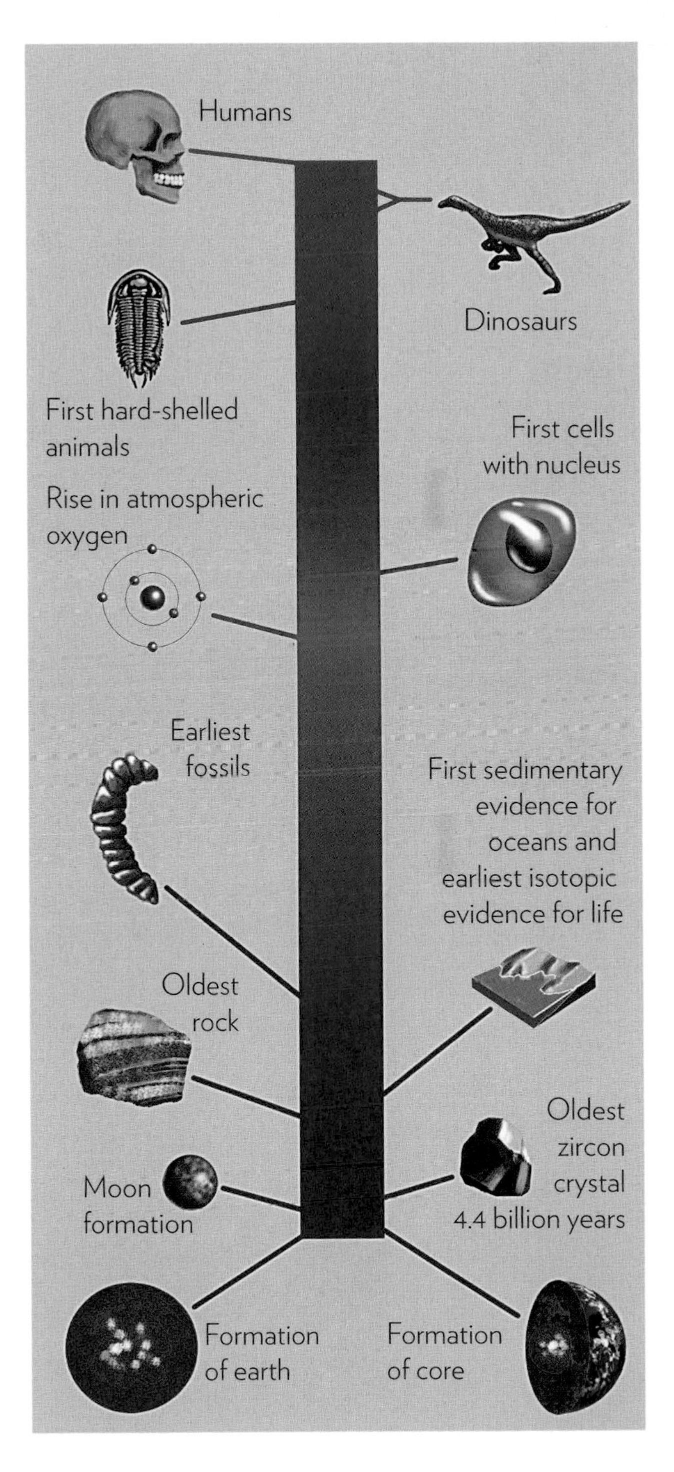

Q:

Why did we end up with a 24-hour day?

Life on Earth is perfectly adapted to the planet's rotation, which turns out to be just fast enough to allow the Sun to heat the surface, but not to burn delicate organic structures like leaves. How did this lucky coincidence come about?

A:

The Earth was probably hit by a large object early in its life, which slowed its rotation just enough to make our day 24 hours. But there are actually several different kinds of "day" on Earth ….

The word *day* is, of course, much older than the science of astronomy and orbital dynamics. For most of human history, a "day" is the period of time between sunrise and sunset—and the other half of the planet's rotation was called night. Everyone knows this!

To scientists, a day is a more complicated concept. For example, the number of seconds between two sunrises (e.g., sunrise on Monday to sunrise on Tuesday) is different from the number of seconds between noon on one day and noon on the next.

When we're talking science, a "solar day" is the time between two noons and is 86,400 seconds long. Scientists use seconds instead of minutes or hours because a second is a scientific unit of time measurement, based on the speed of light.

Scientists can also measure a day based on the movement of a fixed star around the Earth (though of course the Earth itself is moving). This is called a "stellar day." The stellar day is useful because it's the same length all year, while a solar day changes by nearly eight seconds, depending on where Earth is in its orbit around the Sun.

We need all these different types of day because the Earth's orbit around the Sun isn't perfectly circular. The eccentricity of the orbit affects how our planet rotates in relation to the Sun.

The planets in the Solar System rotate because they formed from a vast disc of spinning dust and rock. The laws of motion say that if something is spinning, it has to keep spinning even if you change how far it is from the center of rotation. It's called the law of conservation of angular momentum.

When spinning ice skaters pull in their arms, their distribution of weight changes, but the "amount of spin" doesn't. So they spin faster.

As the dust in the early Solar System started to clump together into planets, the angular momentum of the dust was conserved, and the planets themselves began to spin. The speed of their rotation depended on how much "stuff" accumulated. You can see this by swirling your coffee with a teaspoon, then dusting cocoa on the top. Some cocoa will collect in lumps, and those lumps will start spinning, creating little eddies in the cup. You just made cocoa planets!

More stuff, faster spin. Jupiter's day, for instance, is less than 10 hours long, because the planet is 300 times heavier than Earth.

There is a mystery, though. Earth is the densest and heaviest of the four rocky planets, but our day is nearly the same as the Martian day (which is about 25 hours). But Mars is only one tenth the mass of Earth.

So Earth's day is strangely long. The possible explanation? Something very large, maybe as big as Mars, crashed into Earth early on in our planet's history. It created the moon and changed the rotation of the planet, slowing it down to the current 24-hour day.

Our day continues to change over time. The moon is slowing us down by a few microseconds a year.

A planet's day is affected by many things and can be a source of much mystery to scientists. Venus has one of the oddest days in the Solar System. A Venusian day is 243 Earth days long, and it rotates in the opposite direction to every other planet in the Solar System!

Q: Why doesn't the Earth have more craters?

When we look through a telescope at other rocky planets and moons in the Solar System, we see they have one surface feature in common: craters. Lots and lots of craters, from meteorite impacts. But from space, Earth appears to have no craters at all. Where are our craters?

A: There are lots of craters, but they're blurred and hidden because the Earth is unique in the Solar System. We have two things no other planet has: oceans on the surface and lots of life. And continental drift plays a role, too.

Though we don't yet have a complete answer for how the Solar System was formed, scientists mostly agree that a large disc of matter orbiting the Sun slowly clumped into the eight major planets. But about 1 percent of the material instead formed into trillions—yes, trillions—of rocks, comets, and asteroids.

These objects move throughout the Solar System in all sorts of crazy orbits, and over a long enough period, thousands of them will eventually hit a planet or a moon.

There was even a period in the Earth's early history where the number of "impacts" (rocks hitting something) increased—it's called the Late Heavy Bombardment, and it's why the Moon has so many craters.

So did the Earth just escape getting hit? Not at all—we've been smashed by our share of space rocks. There is strong evidence that a large object, probably a comet, hit what is now Central America and killed off the dinosaurs.

If the Chixulub Impact, as it's known, had hit the Moon, there would be a huge round crater for us to admire. So where's the Chixulub crater on Earth? Why can't we see it?

The crater is there all right, but it's mostly under the ocean. What parts remain on land have been eroded by wind and rain, and the jungle has grown over the top. If you use a satellite and specialized instruments, it's quite easy to see a distinct round geological shape hidden under the familiar coastline of Central America.

Geologists have identified thousands of craters all over Earth. Some of them have lakes in the middle, others are buried under sand dunes, still others can only be detected by the damage they did to the crust deep underground—all surface features have eroded away.

Earth is unique in the Solar System because of our water cycle (liquid oceans that evaporate to create rain on land) and our abundance of life.

Rain and wind erode the distinctive crater walls, smoothing out the jagged peaks you can still see on the Moon. And plants grow too, making it hard for us to spot craters under jungles or grasslands.

Over longer periods of time, the processes of plate tectonics (the way sections of the surface of the Earth move around on top of a liquid interior) jumble and change many surface features. Valleys open up, mountains are pushed into the sky, coastlines sink or rise. All of these things destroy the delicate structure of an impact crater.

But there are still places on Earth where you can visit a well-preserved crater. For instance, the central Australian desert has several craters, such as at Wolf Creek. Because these areas receive very little rainfall, have sparse plant life, and are located far away from tectonic fault lines, the land is rarely disrupted—and so the craters are preserved.

But compared to the craters on the Moon, some of which are millions of years old, even craters like Wolf Creek won't last long. Within a few hundred thousand years, they will literally blow away in the wind and fade away as the surface of the Earth continues to change.

How did Earth get an oxygen-rich atmosphere?

Q:

Strictly speaking, Earth has a nitrogen atmosphere. But there's an awful lot of free oxygen floating around, which is very useful for oxygen-breathing animals like us! But free oxygen is very unusual—how did we end up with so much of it?

A:

Early bacteria evolved a metabolism that released oxygen as a by-product. Every time they ate, they excreted oxygen—which bubbled out of the ocean into the air. But the mix of our atmosphere is always changing ....

Oxygen is a very useful element if you need to move energy around a complex chemical structure like, say, a human being. Oxygen—the word comes from Greek and means "acid maker"—reacts violently with lots of different chemicals, sometimes releasing energy, sometimes binding chemicals together.

But because oxygen is so reactive, it doesn't form as a gas in the atmosphere all by itself. Other life forms need to first make the kind of oxygen humans breathe. Mostly, algae in the oceans make oxygen, though land plants provide a significant portion as well. Oxygen is highly toxic to some life, including certain kinds of bacteria.

Before the emergence of life about 3.5 billion years ago, Earth had an atmosphere mostly made of nitrogen and carbon dioxide. There's so much nitrogen (78 percent today) because nitrogen doesn't react very strongly with many other elements, so elemental nitrogen tends to just seep out of the planet and collect in a gas, held close to us by gravity.

There is a lot of oxygen inside the Earth—it's the most common element in the planet's makeup. But because it reacts so well with other chemicals, most of our oxygen is locked up in compounds called "oxides." Many of our rocks are oxides, including silicon dioxide—which we know as sand. Iron oxide is also very common near the surface, in huge bands of rust. We mine iron oxide and process it to remove the oxygen and get metallic iron for making steel.

Speaking of rust, many human artifacts rust in the open air because the metal in them reacts with the free oxygen. In other words, the oxygen in the air is always looking for a way to react with other chemicals and be removed from the atmosphere.

Oxygen goes back into the air when plants and other photosynthetic life forms expel it after processing carbon dioxide. The plant keeps the carbon and releases the oxygen.

For many millions of years after the evolution of simple photosynthesizers, the oxygen they released was quickly bound back up into rocks and carbon dioxide. But as those primitive organisms—especially a group called cyanobacteria—reproduced and grew more numerous, the rate at which they released oxygen overwhelmed the available "oxygen sinks" on the surface.

Soon, too much oxygen was being produced for it to be bound up in rocks. It began to accumulate in the atmosphere. Scientists call this the Great Oxygenation Event, and it took place roughly 3.5 billion years ago.

Ironically, this was a kind of catastrophic climate change for the life on Earth at the time. Many species of bacteria and single-celled organisms were driven to extinction by the slow "poisoning" of the atmosphere.

Bad luck for those early germs, but good news for complex life. Life based on oxygen has much more "free energy" available, and so we were able to evolve the ability to move around, grow large complex structures like skeletons and eyes, and, most importantly, emerge from the sea to live on land.

It took those cyanobacteria millions of years to make our oxygen. But today there are so many oxygen producers in the biosphere that if all the oxygen disappeared tomorrow, it would take them only about 2,000 years to replenish it to current levels!

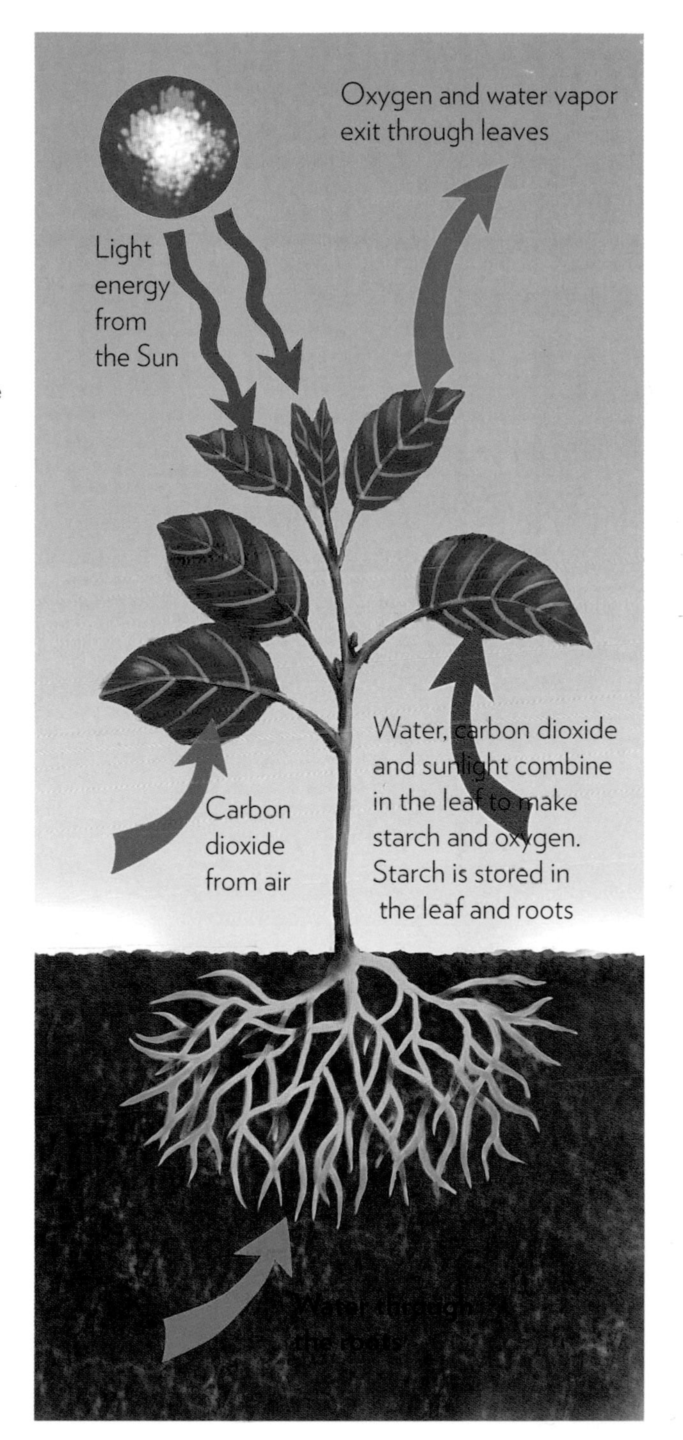

Q: Why is Earth the only planet with a liquid ocean?

As far as we know, Earth is the only planet in the Solar System that supports life. Furthermore, it's the only planet with a liquid ocean on the surface. There are other moons with a lot of water, so why don't they have oceans?

A:

Any water on a planet closer to the Sun than the Earth boils away, and any water farther out freezes solid. It really is as simple as that! Or is it ....?

The question of why Earth's ocean is liquid is a little bit more involved than just saying "it's warm enough."

Earth's orbit around the Sun is in "The Goldilocks Zone." Like the porridge preferred by the anti-heroine of that long-ago fairy tale, Earth is not too hot and not too cold.

But there are other planets—such as Mars—that have summer temperatures warm enough for liquid water, but have no large bodies of water on the surface. Why?

The key is Earth's relatively thick atmosphere. It has a high enough pressure to allow water to exist in all three phases—solid (ice), liquid, and gas (steam). This mix of temperature and pressure is called the "Triple Point" of water.

On Mars, the average temperature is -67°F, though it can get as high as 68°F at the equator in summer. Normally, that would be hot enough for liquid water, but Mars has a very thin atmosphere, so water boils at a much lower temperature than here. All the water has boiled away!

Venus is different—it's closer to the Sun and has an extremely dense carbon-dioxide atmosphere. So it's too hot for water to exist as ice or liquid.

Three planets—Venus, Mars, Earth. One too hot, one too cold, and one just right. They don't call it The Goldilocks Zone for nothing! What's more, Earth actually has a lot more water to make an ocean than Venus or Mars. Again, the thick atmosphere helps, as it traps water molecules on the surface. When water boils on Earth, it gets trapped by the atmosphere. When it cools, the gaseous water turns into first clouds (which are made of billions of liquid water droplets) and then rain, and falls back into the ocean. Or onto land where it runs into rivers ... and then back to the ocean.

Where did we get all that water? From asteroids and comets that crashed into the planet in the first couple billion years of its life. When the Solar System formed, water was created farther out from the Sun than our orbit (the rings of Saturn have a lot of water in them). Later, it fell back in the form of comets. Many of those hit Earth.

Of course, back then there was no life on Earth, so this water delivery service didn't harm anything living.

After smashing into the planet, the water would have boiled into the atmosphere and then rained down onto the hot surface. The cycle repeated over millions of years as the Earth cooled, and eventually the surface reached an ideal temperature to support liquid oceans.

Interestingly, Jupiter's moon Europa has lots of water on its surface. Naturally, it's frozen because Europa is so far from the Sun. But scientists believe there's evidence to show that deep under the ice, the little moon has a liquid ocean. Could life exist in Europa's pitch-black ocean? It's possible!

One of the handy things about water—for life, anyway—is that it freezes from the top down. This means that even at Earth's poles where the ocean is covered in ice hundreds of meters thick, life can still survive. Even large animals like certain species of fish can live under the ice, in their liquid home.

Earth

Temperature just right

Thick atmosphere

Has magnetic field

Venus

Too close to the Sun so too hot

Thick atmosphere

Lacks magnetic field

Mars

Too far from the Sun so too cold

Thin atmosphere

Lacks magnetic field

Q:

Is carbon dioxide the most dangerous greenhouse gas?

There are a number of "greenhouse" gasses, so called because they trap heat and raise the temperature of our atmosphere. CO_2 *gets a lot of press because humans make it, but there are a couple of others worth keeping an eye on.*

A:

The gas that's most effective at trapping heat is water vapor, but humans don't directly affect the amount of water vapor in the atmosphere. But CO_2 and methane could create a feedback loop with disastrous consequences ....

As twenty-first-century civilization grapples with the issue of climate change, most of the conversation revolves around the amount of carbon dioxide in the atmosphere. This is because it's easy to see how humans directly produce CO_2 via industry and transportation. If the question is "Which greenhouse gas is it easiest for us to stop producing?", then the answer is definitely CO_2.

But CO_2 is not necessarily the most dangerous greenhouse gas, if we define "dangerous" as being the gas which traps the most heat. Water vapor traps more heat than CO_2, and there's much more water vapor in the atmosphere. Meanwhile, coming just behind CO_2 in terms of its ability to trap heat is methane.

So why don't we hear more about water vapor and methane? Well, there is quite a lot of discussion about methane, especially coming from farms. Cattle in particular produce a lot of methane, but so do rice paddies (bacteria that live under the rice make methane). Burning anything biological, such as huge tracts of the Amazon rainforest, can also create significant amounts of methane.

Let's deal with water vapor first. Water vapor is extremely effective at trapping heat. It also makes up the majority of greenhouse gas in the atmosphere. But human activity doesn't directly affect the amount of water vapor in the air. The water cycle—the process of evaporation from the ocean, formation of clouds, and rainfall—determines how much water stays in the atmosphere as a gas. But a big part of that equation is how hot the atmosphere is overall. A hotter atmosphere can hold more water. More water increases the heat of the atmosphere. It's what we call a feedback loop.

What starts that loop in the first place? Carbon dioxide. That's even though CO_2 is only 0.06 percent of the overall atmosphere by mass. (The mainstream press usually reports CO_2 in parts-per-million, though, around 393ppm at the time of writing.) Changes of just a few parts-per-million can affect how other gasses get taken up. The feedback loop continues.

Methane is a very dangerous greenhouse gas and is number three on the hit list after water vapor and CO_2. We produce a fair amount of methane, but the real worry is huge reserves of frozen methane locked in the sea floor. If the sea warms enough to melt those deposits, millions upon millions of tons of methane could escape into the atmosphere in only a few years or decades. And that would be catastrophic.

At the end of the day, though, it's not about which gas is the most dangerous based on the principles of chemistry and physics; it's about which gas is the most responsible for the changes in the climate.

Carbon dioxide remains the gas that has changed the most over the period of human industrialization. The level of CO_2 in the atmosphere ultimately affects how much water is in the atmosphere, and it could lead to the release of a lot of methane.

All three greenhouse gasses combine to trap the Sun's heat and increase the overall temperature of the planet. Probably not by enough to kill all life, but certainly by enough to radically change the distribution of plants, animals, deserts, and tropical and temperate regions.

That's what will create problems for humans. And at this stage in our technological development, carbon dioxide is the most manageable of the greenhouse gasses. Time will tell how well we rise to the challenge of climate change.

Solar radiation passes through Earth's atmosphere > Some radiation is reflected by Earth and its atmosphere > Most radiation is absorbed by and warms Earth's surface > Some infrared radiation escapes Earth's atmosphere, into space and some is absorbed and released through greenhouse gases, warming the lower atmosphere and surface > Infrared radiation is released from Earth's surface

Q:

Why are scientists so worried about frozen methane in the Earth's crust?

A special kind of ice called a methane clathrate is found at the bottom of the coldest parts of the ocean. If this ice melts, scientist believe the consequences could be dire for life on Earth.

A:

The buildup of greenhouse gasses in the atmosphere is happening at a more or less manageable rate (if we choose to act). But if methane clathrates in the ocean melt quickly, they could dump lots of methane into the atmosphere at once, with catastrophic results.

Climatologists have identified a number of greenhouse gasses. We all know about carbon dioxide, but another significant gas is methane (see "Is carbon dioxide the most dangerous greenhouse gas?").

Until recently, we thought most atmospheric methane came from biological processes—literally the gas passed by cows! We also knew about methane trapped underground in the form of natural gas, which we're now using as fuel in many places instead of oil.

But there's one other massive source of methane. When methane seeps up out of the crust and encounters freezing water at high pressure—such as at the bottom of the Arctic Ocean—it forms a solid called a methane clathrate.

It looks pretty much like regular ice, except that you can set it on fire very easily. It's pretty weird—the ice burns and melts at the same time, producing a flame and liquid water.

If the methane clathrate melts, the methane trapped in the ice is released as a gas into the atmosphere. The results of that range from "pretty bad" to "total catastrophe."

The pretty bad version of events is when the methane is added to the current levels of greenhouse gas. Lots of methane released very quickly—over a few hundred years—will dramatically boost the rate of global warming and may create what scientists call a runaway greenhouse effect. In other words, we'd be powerless to stop global warming, even if we turned off all our cars and power stations.

If the methane clathrates melt very fast, over a few days, the results are much more dramatic. If the methane is released very quickly, essentially erupting from a deposit on the sea floor, then methane levels in the area will become extreme. If methane levels reach 5 percent of the total atmospheric mix, the air itself becomes explosive. A lightning strike could literally make the air explode, in a region many hundreds of miles across. We could see firestorms beyond the dreams of Hollywood special effects masters.

Following the firestorm, ash and dust would rise up high into the atmosphere and block the Sun, plunging the Earth into a bitter winter. But after the dust settles, the methane remains, trapping heat and leading to a very hot summer. This swing between very cold and very hot would be much worse for humans and our farms than a more constant warming trend.

Fortunately, recent studies of the sea floor indicate that very large methane clathrate deposits are much deeper than initially feared, and the explosive situation seems unlikely to occur. However, monitoring in Siberia has revealed methane levels 100 times higher than normal, coming from melting permafrost.

The other problem is as the methane bubbles up from the ocean floor to the surface, it reacts with oxygen dissolved in the water. Huge sections of the ocean could be depleted of oxygen, which will kill life and result in huge so-called "dead zones."

Studies of ice cores from the Arctic and Antarctica provide many clues to the levels of atmospheric methane in the past. Scientists now believe that spikes in methane match up with some of the big extinction events of millions of years ago.

In short, we'd rather all that methane stay frozen.

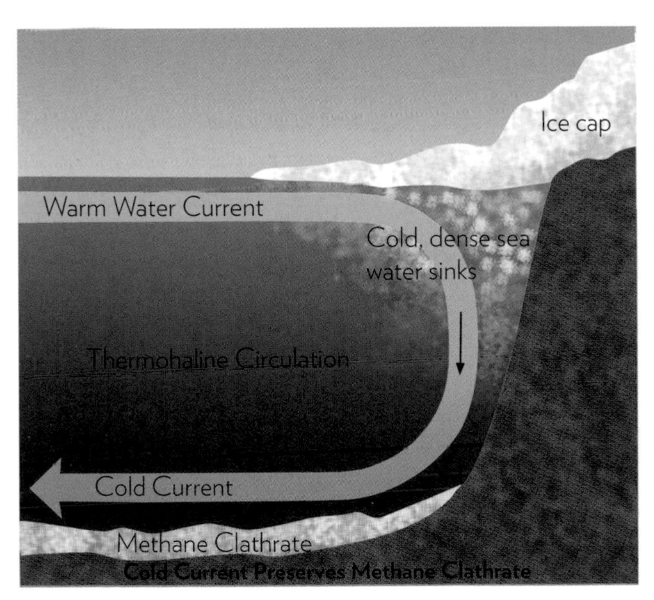

Cold Current Preserves Methane Clathrate

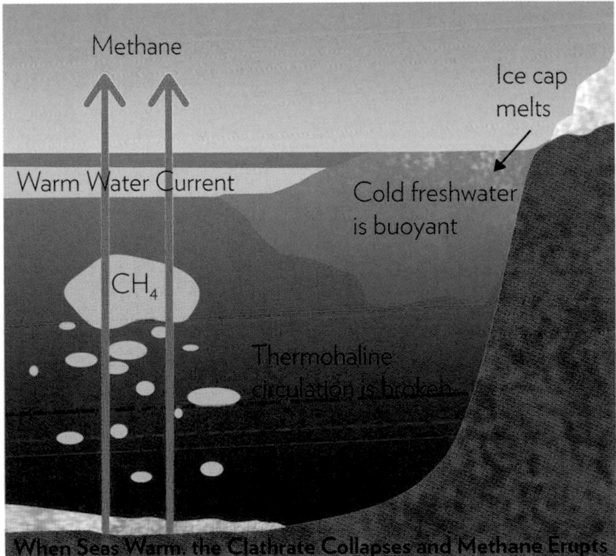

When Seas Warm, the Clathrate Collapses and Methane Erupts

Why is the chemistry of the ocean so important?

In any discussion about life on Earth, climate change, or the health of the biosphere, it isn't long before the chemistry of the oceans comes up. Acidity, salinity, and the amount of oxygen in the water are all vital indicators of the health of our planet.

A: All life on Earth is, in some way, dependent on the oceans. The current biosphere is hugely dependant on the exact mix of chemicals in the water, including the overall acidity of the ocean and the amount of dissolved oxygen. If these levels change, life as we know it could be severely disrupted.

When it comes to conservation and ecology, there's a lot of focus on the health of the ocean. This might seem odd for a bunch of land-dwellers like humans. Sure, we eat fish from the ocean, and we want to manage fish supplies, but why do we care so much about the chemical balance in the sea?

While there are some "extreme" bacteria and other micro-organisms that live deep in the crust and seem to be independent of the ocean, every other life form on the planet owes its existence to the sea. The ocean is the ultimate starting point for all the food webs and other biological processes that make complex life possible—even on land.

Tiny organisms called phytoplankton are the foundation of the biosphere's ability to feed itself. Slightly larger predatory plankton eat the phytoplankton, and progressively larger creatures eat each other. What's more, a considerable amount of the phytoplankton is algae, and this algae produces 50 percent of the oxygen in both the sea and the atmosphere.

These tiny plants and animals are very sensitive to the chemical balance of the ocean. Because seawater contains lots of salt and trace amounts of other substances like calcium and magnesium, not all seawater is created equal.

In some parts of the world, the ocean is saltier, has higher calcium levels, is more acidic, or exhibits many other characteristics. Life in those parts of the ocean varies. There are even "deserts" in the ocean, where a lack of oxygen and other essential minerals means life cannot exist.

All these differences are normal and natural, and they change slowly over long time periods. Usually, life has enough time to adapt. If an ocean warms, the coral reefs will retreat toward cooler water and stop when they reach water that's too cold.

If the change in conditions occurs over many thousands of years, the life in the ocean has plenty of time to respond. Problems arise when ocean conditions change rapidly. Coral grows fast, but not *that* fast.

Another really important aspect of ocean health is acidity. To a human, the differences typically seen in ocean acidity—its so-called pH level—are imperceptible. A more acidic ocean won't burn your skin when you go for a swim.

But a more acidic ocean does affect an organism's ability to build a calcium-carbonate (chalk) skeleton. Coral is the most famous creature to use calcium-carbonate, but almost all the important plankton species use it, too. If the ocean is too acidic, these skeletons can't be formed. If the plankton can't form a skeleton, it doesn't develop properly, and populations crash. Even worse, existing corals and planktons may find their skeletons dissolving!

When a region of the ocean loses its plankton population, other life dies, too. Tiny fish and filter-feeders like barnacles and jellyfish die, and then large creatures follow.

What's more, without plankton, oxygen levels drop. Without dissolved oxygen in the water, fish can't breathe.

How do the oceans get more acidic? It's the fault of atmospheric carbon dioxide. When CO_2 dissolves into the ocean, it reacts with water to form carbonic acid—more CO_2, more acid, and more dramatic change in pH level.

Cutting CO_2 levels will stop the further acidification, and help stabilize the ocean.

375 parts per million
33.8°(C)

450-500 parts per million
35.6°(C)

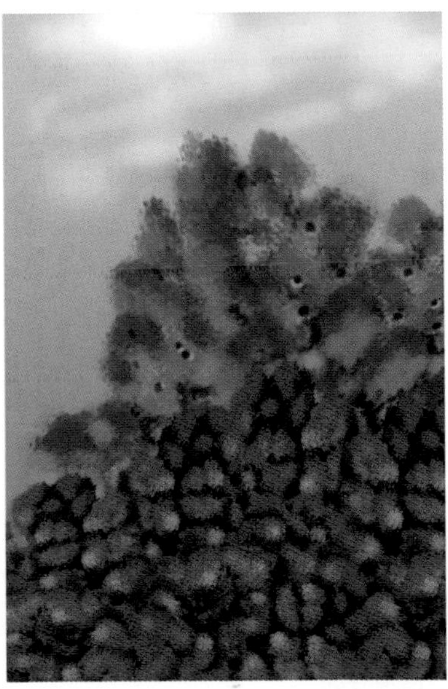
> 500 parts per million
> 37.4°(C)

Q:

Could an earthquake ever sink a whole country?

Years of awesome disaster movies and sci-fi novels have predicted devastating results from super-powerful earthquakes. Entire countries sinking beneath the sea! Atlantis! Giant mountains appearing like magic! The truth is less spectacular ... but no less destructive.

A:

Even the largest earthquake ever recorded didn't change the coastline of Chile. In fact, since most really big earthquakes are "megathrust" quakes, they're more likely to lift a land mass farther *out* of the sea. Volcanoes, though, are a different story ....

Everyone who lives in the state of California or the Japanese megacity of Tokyo has heard of the mythical "Big One"—an anticipated superquake that will cause the land where they live to break off and sink into the ocean.

Sadly for the writers of disaster movies, earthquakes—especially the super-destructive kinds—don't work like this at all. Still, a really big quake can shift huge tracts of land around. After all, tectonic activity is what forms mountain ranges, where the land is pushed together and upward, like a crease on a bed sheet. And a quake near a coast will cause a tsunami, which can flood the land and make it seem like the country has sunk into the sea. But the water will recede, and sea levels won't change.

There are several different kinds of earthquakes. Seismologists tell them apart by how the land on either side of the fault moves. Some quakes occur where one tectonic plate slides along another, catching on the edge and then releasing all that pent-up friction at once.

Other quakes occur where one plate is being forced underneath another in a head-on collision. These are called megathrust quakes and, based on current records, are the most powerful and destructive.

But ironically, megathrust quakes actually *lift* the land. There was a very powerful megathrust quake near the Greek island of Crete in 365 A.D. Though it destroyed nearly all the towns and settlements on the island, it lifted the land nearly 30 feet (9m) higher. Quite the opposite of sinking!

That's not to say that an event in the Earth's crust couldn't "sink" a large island. Volcanoes can have incredibly massive effects on the land around them.

We've been lucky in the last couple thousand years, with very few truly enormous volcanic eruptions. There are some standout exceptions, though. In 1883, a volcanic Indonesian island called Krakatoa erupted. The force of the explosion was enough to destroy two thirds of the island. To anyone passing by, when the dust and ash cleared it would have seemed as if the island had sunk beneath the sea. In fact, the rock and dirt was hurled outward and the sea rushed in to fill the crater—or *caldera,* as it's known.

The process of plate tectonics and continental drift is a very gradual one. Without human cities, roads, power grids, and other fragile infrastructure, the damage caused to the surface by an earthquake is actually pretty mild. Trees fall, rivers change course, the land floods briefly, but life bounces back.

Despite the tragic loss of life and the huge cost, humans, too, recover quickly from earthquakes—especially in developed nations. In scientific terms, quakes are of much less concern than the eruption of a supervolcano, which has the power to effectively sterilize a huge swath of land around it.

However, the biosphere relies on both quakes and volcanoes to create fertile land. Plants grow well in volcanic soils, and earthquakes can bring water to arid areas and slowly build mountain ranges that stimulate rainfall.

Many scientists believe that without active plate tectonics, Earth would not have such a rich abundance of life. Far from destroying our world, earthquakes and volcanoes may actually help make it.

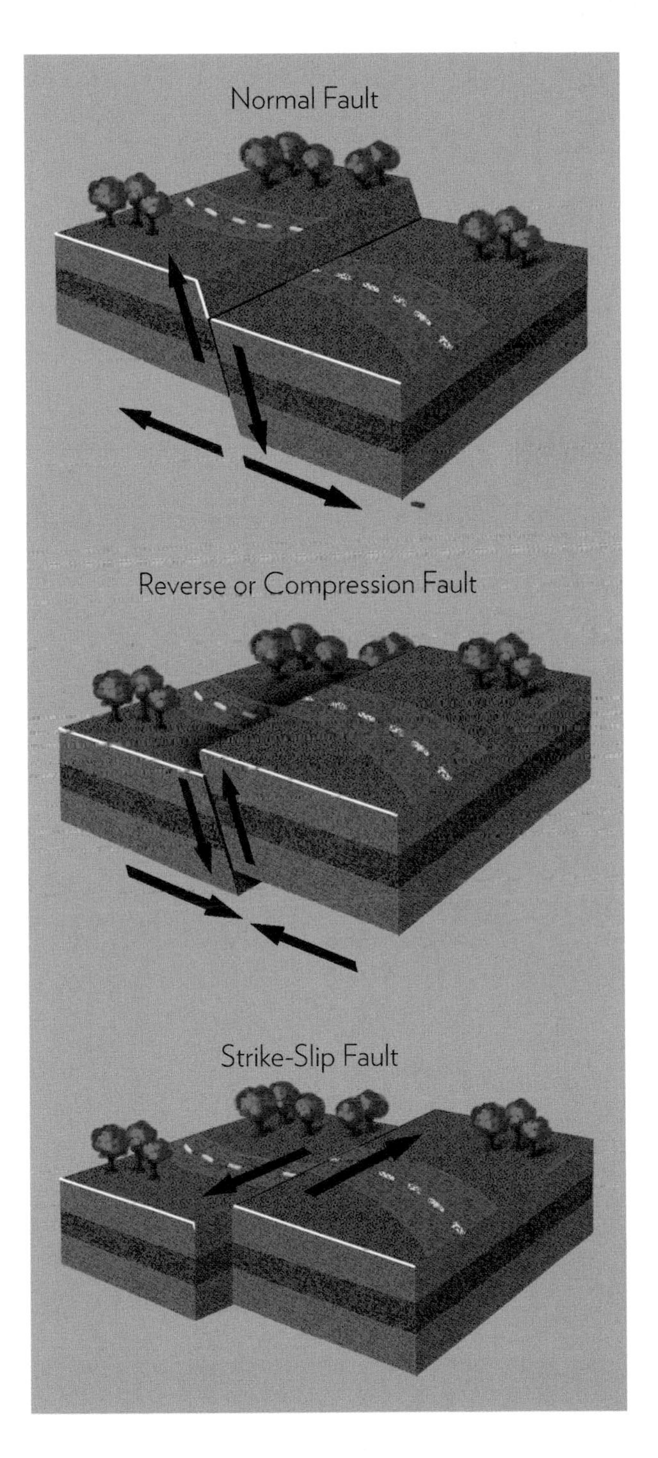

Q:

Should I be scared of supervolcanoes?

A supervolcano is a popular term for an extremely large eruption. Far from the iconic conical mountain of a typical volcano, a supervolcano can be many miles wide. Modern humans have never seen a supervolcano eruption, but the evidence is all around us.

A:

There are a few supervolcanoes, dormant for now, dotted around the world. One lies underneath Yellowstone National Park in the United States, and the whole island of Iceland is another. If either were to erupt massively, it would almost definitely mean an end to our civilization.

As the science of geology became more sophisticated throughout the twentieth century, scientists learned more and more about how the ground under our feet formed.

What became obvious is that there are places in the world where land has formed from huge upwellings of magma and lava, vast tracts of land created (geologically speaking) almost instantaneously by volcanic eruptions on a scale never before imagined.

While a typical volcano like Mount St. Helens might produce a caldera (or crater) as much as a mile wide, the caldera of a supervolcano can be hundreds of miles wide. The explosion of Mount St. Helens pumped 0.2 cubic miles (32m) of ash into the air. A supervolcano eruption like Yellowstone or Toba in Indonesia could eject as much as 240 cubic miles (386km) of ash.

How would that affect us? Pretty badly. Much of the ash would rise into the upper atmosphere and shroud the planet, blocking sunlight. More ash would rain down across the globe, smothering the land it covered, killing plants and fouling rivers and lakes. The planet would cool rapidly, first falling into what's called a "volcanic winter" and perhaps even entering a short ice age lasting 1,000 years or more. Eventually the ash and dust would fall out of the atmosphere and the planet would recover.

It's unlikely much of human civilization could survive such an eruption. Without sunlight, our crops would fail and billions would die of famine. Ash would ruin our arable land and poison our water. Some pockets of people would probably survive in bunkers or by scavenging the ruins, and we'd slowly rebuild. But essentially, the result would be much the same as a large-scale exchange of nuclear weapons. There would be less radiation, but more global cooling and widespread collapse of the biosphere.

Why do scientists think this? Because there's considerable evidence to suggest it's happened before. Roughly 75,000 years ago, the Toba supervolcano on the Indonesian island of Sumatra erupted and covered the whole of South Asia in more than a foot of ash. At roughly the same time, anthropologists believe there was a "genetic bottleneck" in the human species and that our population fell to as few as 10,000 people.

Is the bottleneck the direct result of the Toba eruption? It's not easy to prove, but the coincidence is compelling. The eruption would have led to severe drought in the tropics and the loss of food sources. Humans would have needed to adapt to new environments and figure out new ways of surviving on a much cooler planet. An ice age followed the eruption almost immediately.

Should you be scared of supervolcanoes, though? There's not much point. At our current level of technology, we have absolutely no way to prevent an eruption. Organizations like the United States Geological Survey monitor so-called "hotspots" like Yellowstone for signs of increased volcanic activity. The USGS does consider Yellowstone a "high threat system" when it comes to volcanic and seismic trouble spots.

We just have to hope luck stays on our side, and these massive holes in the surface of the Earth stay crusted over for a few thousand more years—until we have the technology to deal with them.

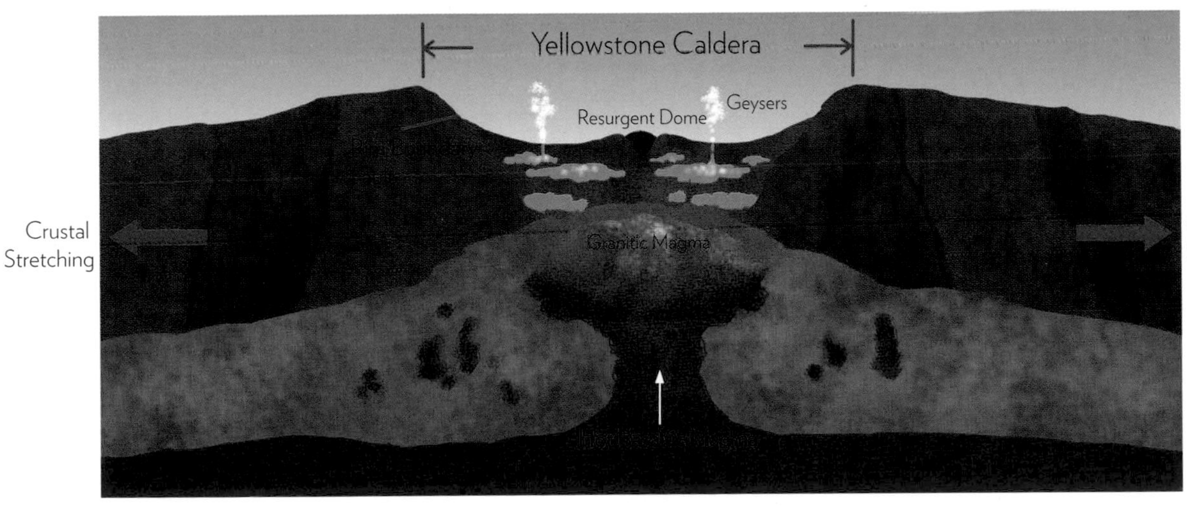

Q:

Why are there so many different types of minerals?

The geology of the Earth is amazingly complex, with thousands of different kinds of rocks and crystals. Some are distinguishable only by experts, while others exhibit astonishing variety in form and beauty. Where does all this come from?

A:

The explosive power of an ancient supernova created hundreds of heavy elements, many of which ended up in the Earth and reacted with each other to form our minerals.

When we look at the Earth on a large scale, scientists say things like "we have a silicate crust and a solid iron core." This can make the planet's structure seem fairly basic and simple, but in fact as soon as you start actually digging, you'll find a huge variety of minerals—about 4,660 different formally identified types.

All the stuff in the universe is divided up into different elements. An element is a single atom with a particular number of protons in its nucleus. Hydrogen, the most basic element, has only one proton. Helium has two. Carbon has six, and oxygen eight. Each element also has a collection of electrons, and it's these electrons that allow atoms to join together into molecules. (For more, see the Chemistry part of this book.) Join millions of molecules together and interesting crystalline structures start to emerge. Once billions of molecules are built up into a crystal lattice, you have a mineral.

A "crystal lattice" is a pattern in the way the atoms are joined together, in triangles or hexagons or something more complex. Under a microscope, even plain grey rocks show a crystal structure.

Water is not considered a mineral (because it's liquid), but natural ice is. Ice cubes made in a freezer are not a mineral, as they are artificial, but snow is. Bone itself is not a mineral because it is grown biologically, though it contains minerals that the animal has eaten.

By weight, the Earth is mostly made of iron, oxygen, silicon, and magnesium. But all 98 naturally occurring elements are in the mix somewhere, sometimes in tiny trace amounts—bismuth, for instance, only appears as one atom in every billion. These elements came from the supernova of an ancient star. When the star exploded, basic elements like hydrogen, helium, and lithium fused to create a mix of heavier elements, including gold, silver, tin, uranium, and more.

The way these elements react with each other to create different crystals gives us the huge variety we find in the crust. The most common mineral you can find on the surface of Earth is quartz, made of silicon and oxygen. Throughout the whole crust, though, including underground, the most common mineral is feldspar, which is made of aluminum, silicon, oxygen, and either sodium, potassium, or calcium.

Because the Earth formed from a molten ball of matter, for many millions of years, all the elements inside the Earth were free to circulate and react with each other to produce this huge variety of minerals. As the planet cooled, crystals solidified into the minerals we know today.

There seems to be almost no end to the variety of shape and color of minerals. From the lustrous shine of gold to the geometric regularity of some quartzes, to more exotic things like the black columns of Hübernite, which glow red when you shine a light behind them.

What's fascinating about minerals is that no matter how exotic they look, they are made of only a few elements bonded together. Corundum, which is the mineral that makes sapphires, rubies, and emeralds, is simply Al_2O_3 (a type of aluminum oxide).

New minerals continue to be discovered to this day, and who knows what exotic things are waiting for us on planets like Mars?

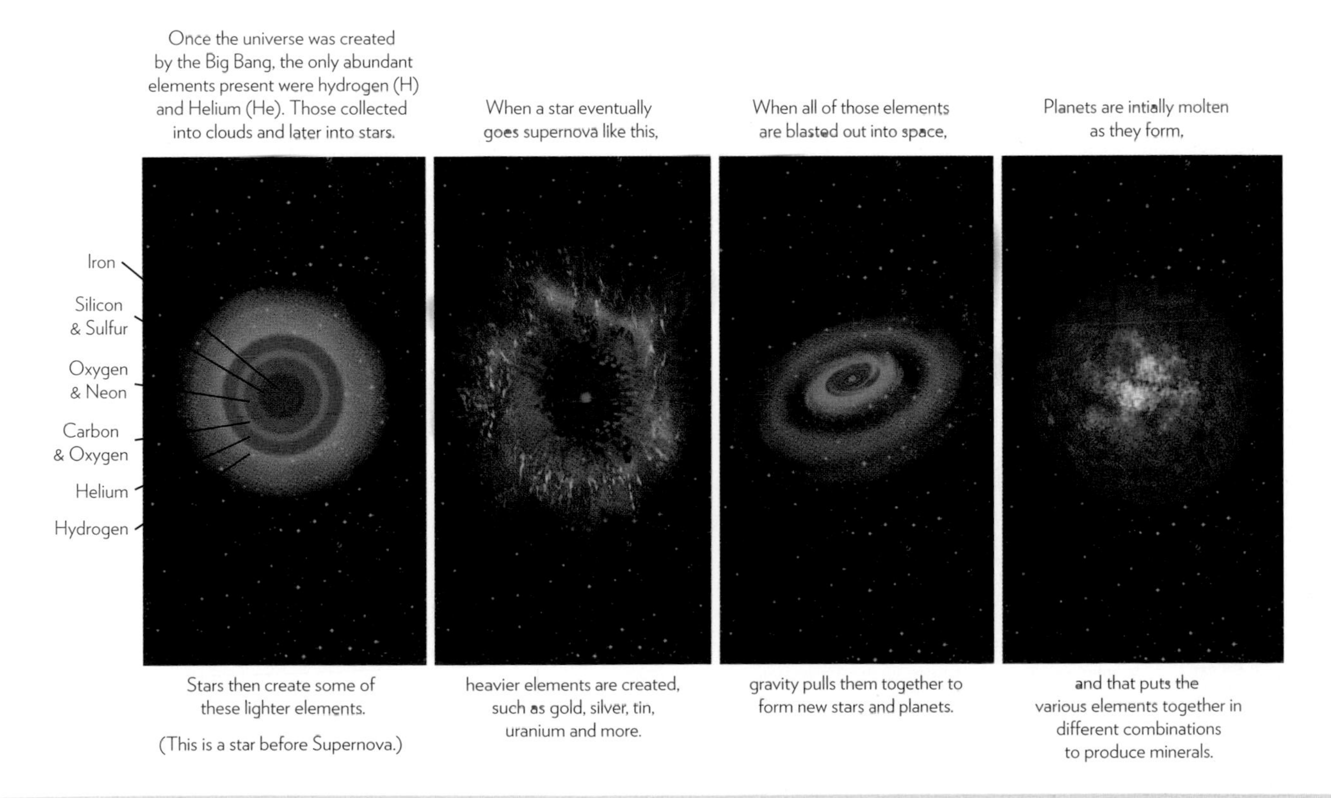

Q:

Are gold and diamonds good for anything besides jewelry?

Humans spend a huge amount of time and money looking for gold and diamonds. We even fight wars over them. Sure, most people agree these minerals are pretty and desirable, but are they actually good for anything more than looking pretty?

A:

We haven't been wasting our time: it turns out gold and diamonds both have unique properties that make them useful for all kinds of things, from electronics to medicine and space travel. That said, while gold is genuinely rare, diamonds can be made cheaply ....

For as long as humans have been able to work metals, we've been obsessed with gold. This ultra-rare, soft, shiny metal does something to our brains—we lust after it, we'll travel the world looking for it, and risk our lives to dig it out of the ground.

Gold is very rare on Earth, occurring only 21 times in every billion atoms on the planet. Put it this way: if humans could be made of gold, there would be only 147 golden people among our population of seven billion.

Diamonds, on the other hand, are made of carbon. Yes, the same stuff as trees and people and pencils. There are no tricky trace elements in diamond, but natural diamonds are found only in areas where the crust has been subjected to tremendous heat and pressure. Diamond forms when carbon is crushed and squeezed and its atoms are forced into a particular crystal lattice that sort of looks like little pyramids.

Diamond is very hard, once considered the hardest naturally occurring substance—though scientists have discovered other ultra-rare forms of carbon that are a little harder. That makes it very useful in industrial applications, especially super-fine drills. Diamond dust is also used for grinding, and tiny specks of diamond can be embedded in disc sanders.

Because diamonds are only made of carbon, we can actually make them artificially. These artificial diamonds have no value as gemstones, because they're too perfect—they have no interesting coloring or exceptional characteristics.

We mine about 60,000 pounds of diamonds from the Earth every year and make another 240,000 pounds. Eighty percent of natural diamonds are used for industrial purposes, and the remainder are sold as gems. Very expensive gems. Why so expensive? The simple answer is because people are prepared to pay for them, so companies charge what they want. There is no special reason for diamonds to be the most expensive type of gemstone.

Industrial Diamond Cutting Blade

Gold, on the other hand, has many applications beyond jewelry. Gold is extremely good at reflecting light and heat, so it's used for insulation on very sensitive electronics, such as satellites and space probes. Ever seen a NASA astronaut with a gold-colored visor on his suit? The visor is actually coated in gold to reflect sunlight and prevent overheating.

Gold can be worked in such a way to become entirely transparent. It's then layered onto aircraft windows and hooked up to a heating system. The system pumps heat through the gold layer—it's very good at conducting heat—and prevents the window from icing over.

High-grade electronics also use gold to conduct electricity. If you buy expensive cables for your home theater, it's likely the ends will be coated in gold.

Gold is also useful in medicine, for testing for the presence of viruses. And of course gold is an excellent replacement for human teeth.

And then there's money. People still invest in gold and track its global price.

How much rarer is gold than diamonds? A lot: as of 2013, only 192,000 tons of gold have been mined in the whole of human history.

Gold Plated Coaxial Male Pin

What would happen if we desalinated the entire ocean?

All life on Earth depends on water, and all life on land depends on being able to drink freshwater. But less than 1 percent of the Earth's entire water supply is fresh and liquid. We have the technology to desalinate seawater, so what would happen if we took all the salt out of the ocean?

A:

Desalinating the ocean even partially would be catastrophic for all sea life, which depends on the salt in the water to survive. Life originally evolved in mineral-rich, salty water. Land-dwellers who depend on freshwater are the exception, not the rule ....

One of the great ironies of living on land is that you need water to live ... but most of the planet's water is undrinkable because of the salt and other mineral content. Only 2 percent of the Earth's water is salt-free, and three quarters of that is locked up in the polar ice caps.

That leaves just 0.5 percent of our total water in liquid, drinkable form. The good news is, that still represents many billions of gallons. The bad news is that the global human population is now so huge, real pressures are mounting on that water supply.

We do have the technology to desalinate seawater. The process is surprisingly simple: we pump water through a processing plant that uses either membranes or differences in pressure to remove the salt and other minerals. Saltwater goes in, freshwater comes out (and the dry salt goes back in the ocean).

Today, we don't have the technology to desalinate the entire ocean and convert all the planet's water to fresh, but we certainly do have the scientific knowledge to do it—it's just a matter of building a *lot* of pumps. But changing the salinity of the oceans could, ironically, kill us all.

Phytoplankton are the foundation of all food webs on the planet, and these microscopic plants also produce half of our oxygen. What's more, they've evolved to live in a very salty ocean. The salt in the sea affects the way energy and food can move in and out of their cells. Single-celled phytoplankton feel this the most strongly, but even large animals like fish are sensitive to changes in salinity.

Sometimes the amount of salt in a particular part of the ocean will drop, especially near the outflows of massive rivers like the Amazon. If the salt level in seawater drops too low, creatures in the area risk going into "osmotic shock." The chemistry of the water affects how water will move in and out of their cells. Too little salt, and cells will fill up with water and even rupture. Phytoplankton can literally explode if there's not enough salt in the water.

Saltwater fish have evolved to absorb lots of water to "flush" salt from their bodies. In freshwater, they become waterlogged, their internal membranes and organs are damaged, and they die.

This is not to say we should stop using water desalination plants. In fact, desalination is probably essential to the long-term survival of our civilization.

Throughout history, droughts and disruptions to freshwater supplies have emptied cities, destroyed nations, and killed millions. Desalination can end our dependence on fragile freshwater sources.

However, these desalination plants require quite a lot of energy to run. This is usually supplied via electricity, and critics of desalination say the system uses too much power to be sustainable. But many desalination plants are built in conjunction with wind farms or solar panel farms to offset their electricity use.

Recent estimates suggest that converting to desalination plants away from freshwater dams would add only 10 percent to the electricity usage of a country like the United States. Split across the population, that's about as much power as running an extra refrigerator per person.

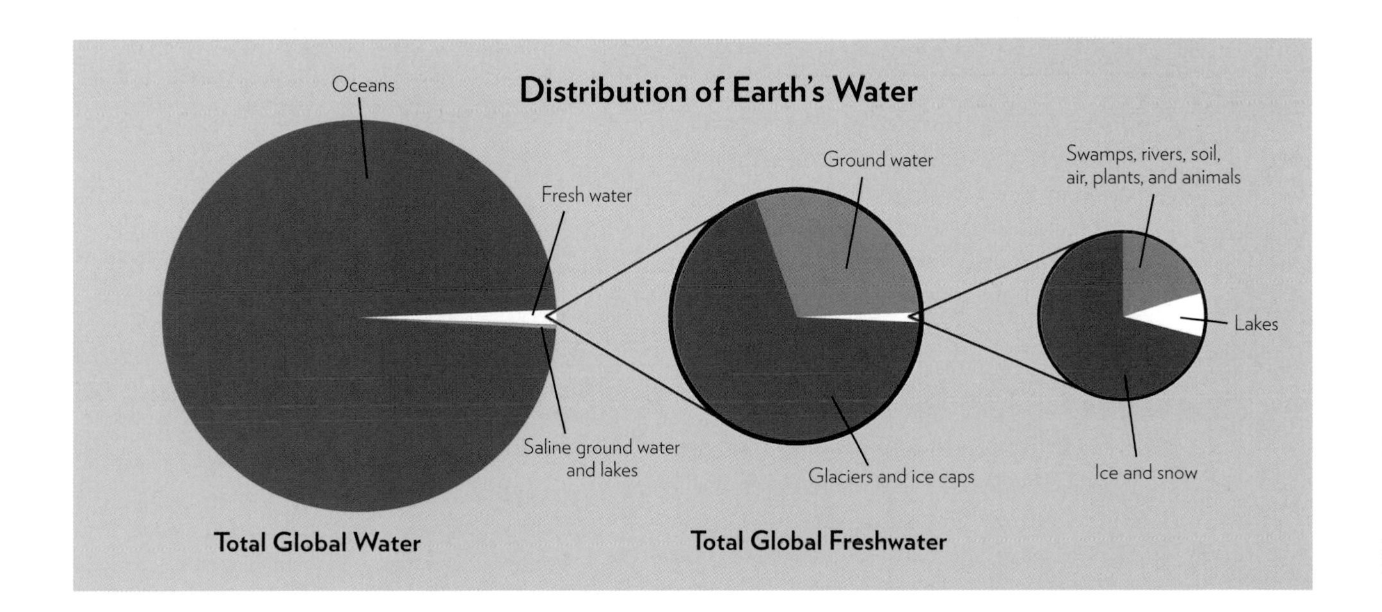

Q:

What would happen if the ice caps completely melted?

More than three quarters of the Earth's freshwater is locked up in the ice caps. If all that ice suddenly melted, the results would be catastrophic ... but also unexpected.

A:

The Antarctic ice would cause a huge rise in sea levels, but also massive earthquakes. And the melting of Greenland's ice could, strangely, freeze Europe ....

Climate change scientists have been warning the world for some time now that one of the effects of global warming will be a rise in sea levels. Current models suggest melting glaciers and ice from around the edge of the polar ice sheets could add as much as 37 inches (94cm) to average sea level. That could cause considerable damage in low-lying areas and make many coastal cities more vulnerable to storms and high tides.

If the whole of the northern ice sheet melted, it wouldn't make that much difference to that figure. That's because there's no land at the North Pole—the ice is floating on water. And as we know, if we let the ice in our drink melt, it doesn't cause the drink to overflow.

Antarctica is another matter entirely. The southern ice sheet is much bigger—it's 7,000 feet (2,133m) thick and contains 90 percent of the world's ice. It's also sitting on top of an entire continent. If that ice melts, it will add a bit more than 37 inches (94cm) to the ocean. About 200 *feet* (60m) more.

Greenland has the next largest ice sheet, enough to raise the oceans by a further 20 feet (6m) should it melt entirely.

But there are other consequences of a massive melting event that go beyond sea level rise.

Antarctica is a very strange place geologically. The ice on the continent is so thick and heavy, it's pressed the surface of the Earth inward, a little like a dent in a Ping-Pong ball. If the ice melted and flowed to the ocean, the pressure on the land would be removed and the crust would pop back out again. The whole world could be wracked by massive earthquakes. There are also active volcanoes in Antarctica that could erupt if seismic activity nearby increased.

If the ice caps are melting, that implies the ocean is hotter overall. More heat in the ocean provides more energy for superstorms like hurricanes and cyclones. While there might be fewer storms per season overall, the storms that do form could be much more powerful than any we've experienced so far. Typhoon Haiyan, which struck the Philippines in 2013, may be just the first of a new age of superstorms.

It's just one of the side effects that demonstrate how complex the issues surrounding climate change really are. It's also why we use the term "climate change" rather than "global warming"—yes, the whole system is getting hotter overall, but local results might be the opposite, at least for many years.

Ice has one more important role in our climate: its shiny whiteness reflects a lot of sunlight. Reflectivity of a planet is called its "albedo," and if Earth maintains a high albedo it stays colder as more sunlight is bounced off into space. Less ice means lower albedo, which means more sunlight absorbed, which means higher temperatures ... thus creating a "feedback loop."

At this stage, it seems unlikely the ice sheets of Antarctica or Greenland will melt in any time period shorter than many thousands of years. Even so, over the next hundred years the partial melting we're already seeing will raise sea levels and have damaging repercussions for our civilization. And also for the ecosystems that have adapted to these huge expanses of ice at the top and bottom of our world.

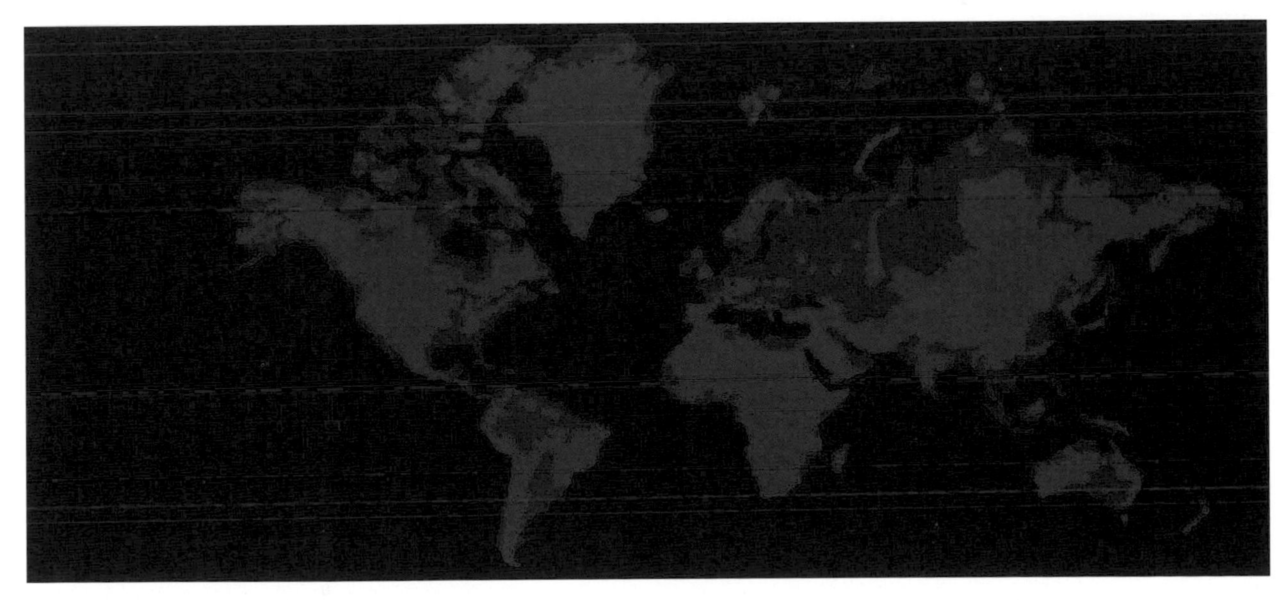

The World if the Icecaps Melted

Current global water — Underwater if ice caps melt — New land formations after melt

Q:

Why does a hurricane have an eye, and why is it so calm?

The massive, spiral-shaped storms we know as hurricanes, cyclones, and typhoons are among the most powerful forces on the planet. Yet in the middle of the strongest storms, a circular region many miles across has blue skies and calm winds. Why?

A:

For reasons not yet fully understood, when a hurricane gets powerful enough, air is forced down through the center of the system, creating the calm eye. But this can be the most dangerous part of the storm ....

For all of our technological cleverness and dominance of the biosphere, humans are still very much at the mercy of nature's most powerful forces. Among these are hurricanes, cyclones, and typhoons.

Despite decades of detailed study, the exact reasons for why hurricanes form isn't yet fully understood. We do know that areas of low air pressure—called depressions—can sometimes join up and begin circling around a central point. As the power of this system ramps up it creates a positive feedback loop, making the storm stronger.

At some point in this process, a region in the center called the "eye wall" becomes especially powerful, with winds rotating faster than in the rest of the storm.

While a hurricane resembles water spiraling down a plug hole, it actually works in more or less the opposite way: air is being sucked in from the sides where pressure is higher, then hurled into the upper atmosphere where it spreads back out in a spiral pattern.

But, when the hurricane becomes powerful enough, it starts sucking air *down* through the center. Why this happens isn't fully understood, and there are hundreds of theories.

This downward force is enough to create a region of incredibly low pressure, as much as 15 percent less than normal, and a circular area of calm and blue skies.

Don't be fooled, though—the eye can still be very dangerous. You might think that ships trapped in a hurricane should make for the eye and stay there until the storm blows out. But in the eye, massive waves as high as 130 feet crash together and come from random directions.

Worse still, while the eye is the calmest part of the storm, the eye wall is the most violent. Many people lose their lives in hurricanes because when the eye passes over, they emerge from their shelters—the calm of the eye may be the first blue skies they've seen in over a week. But if they remain out too long, they can be caught off guard by the opposite side of the eye. Calm weather can turn to powerful winds in moments.

Meteorologists use the eye as an indicator of the power of the storm system. The most powerful and destructive hurricanes have very large, calm eyes anywhere from 3 to 60 miles across. Some storms have very skinny eyes called pinhole eyes, and these can sometimes form sloped walls like a sports stadium—and are a similar size.

Some eyes can be filled with clouds, or even be hidden within the storm. Scientists spot these using weather radar and infrared cameras.

Every hurricane season, brave researchers called "storm chasers" risk their lives flying specialized aircraft inside hurricanes to take measurements. They'll pass through the violent winds of the eye wall in specially designed aircraft and armored trucks to see how the structure of the eye works. Weirdly, some storms can even form hexagonal eyes. We've also seen this phenomenon at the poles of Saturn, where storms bigger than the whole Earth form strangely beautiful geometric patterns.

All this power comes from a simple drop in atmospheric pressure of just a few percent. But the consequences can change people's lives forever.

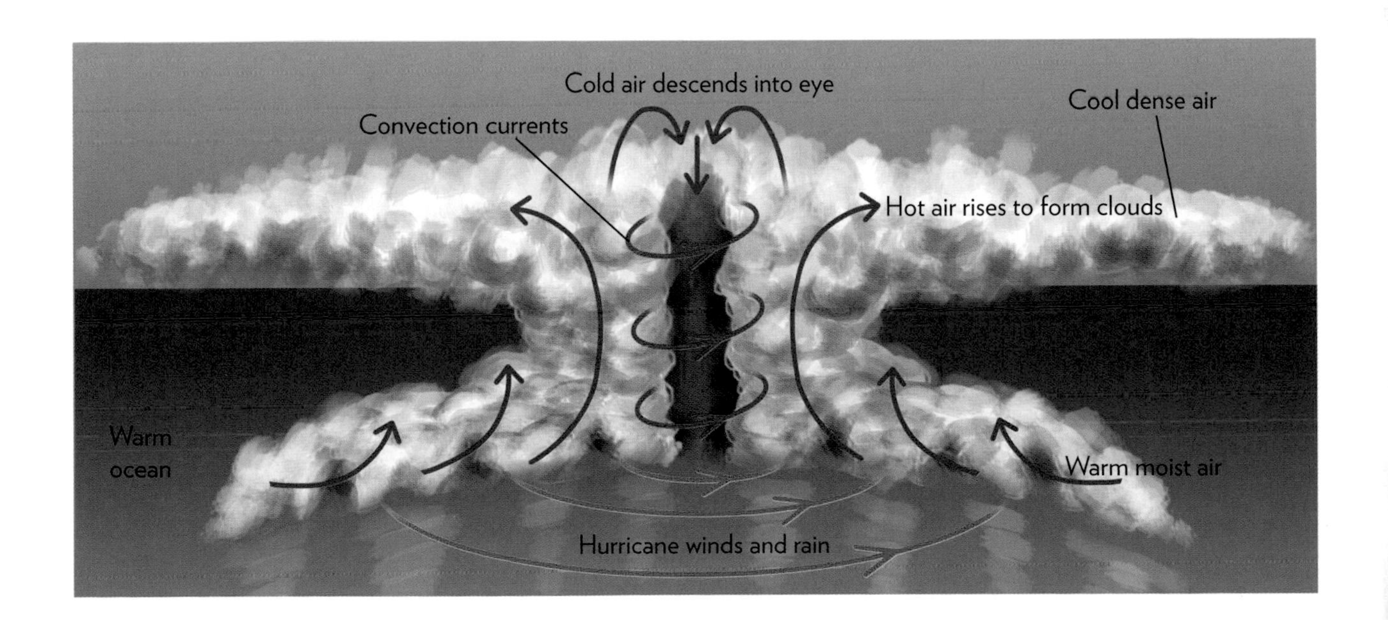

Q: Could we ever use up all the oxygen in the atmosphere?

With a population of seven billion and countless fires, furnaces, and other oxygen-burning technologies, could humans ever inadvertently use up all the oxygen? How secure is our oxygen supply, anyway?

A: While theoretically humans could kill all oxygen producers and consume all available oxygen, this is currently well beyond our capability. But we don't need to use ***all*** the oxygen to make the atmosphere unbreathable ....

The amount of free oxygen gas in the Earth's atmosphere is very unusual (see "How did Earth get an oxygen-rich atmosphere?" for more). If aliens ever scan the planet, they would use the existence of oxygen as evidence Earth supports life.

Oxygen is produced through biological processes. Photosynthetic organisms consume carbon dioxide and release oxygen. About half of our oxygen comes from phytoplankton in the ocean—tiny microscopic plants, mostly types of algae. The rest comes from other ocean sources, and about 30 percent comes from land plants.

Humans are pretty good at destroying vast areas of plant life, but we rarely leave the land we clear empty. Usually we plant other crops that, while not as good at producing oxygen as a mature rainforest, do still release the gas into the atmosphere.

Today, the atmosphere is about 20 percent oxygen. Oxygen is the most common elements on and in the planet. When we burn something or breathe, the oxygen isn't destroyed, it just combines with other elements—usually carbon—to form a new molecule.

In a worst-case scenario where global oxygen levels start dropping dramatically, humans could build machines to generate oxygen from CO_2 and even from rocks. In fact, NASA scientists are currently developing systems to mine rocks on Mars and the Moon for oxygen—the idea being that a spacecraft visiting either place could make its own liquid oxygen rocket fuel for a return journey.

While oxygen is vital to all large life forms on Earth (there are types of bacteria that don't need oxygen, but they still need water), to a chemist oxygen is a dangerous, toxic substance. It ruins samples of other elements by reacting with them, it kills many types of microscopic life, and in high enough concentrations it's incredibly explosive.

The current atmosphere has just the right balance of gases to allow life to extract energy by reacting oxygen in its cells. Humans also use oxygen to start fires—the most fundamental source of power for our civilization. All of our power sources rely on oxygen to some extent—especially if you consider the refined metal parts that must be made in blast furnaces.

If we do end up in a situation where oxygen levels are dropping, we don't need to use up all of it to cause major problems.

Currently, the atmosphere—and every breath you take—is about 20.95 percent oxygen. The various health and safety standards around the world warn against working in an environment where the oxygen has dropped below 19.5 percent.

This doesn't give a lot of wiggle room for humans to mess around with atmospheric oxygen levels. Fortunately, the sheer mass of the entire atmosphere is so huge, it's hard to come up with a scenario where we could reduce levels by 1.5 percent globally.

Management of our atmosphere will be an ongoing concern. We know from ice cores and evidence in rocks that the mix of gasses can change dramatically over long periods of time. The challenge will be to keep that mix adjusted just right for the conditions we want to live under.

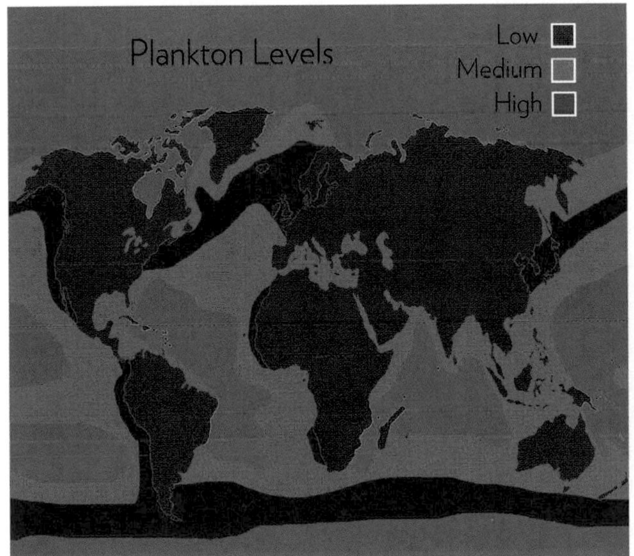

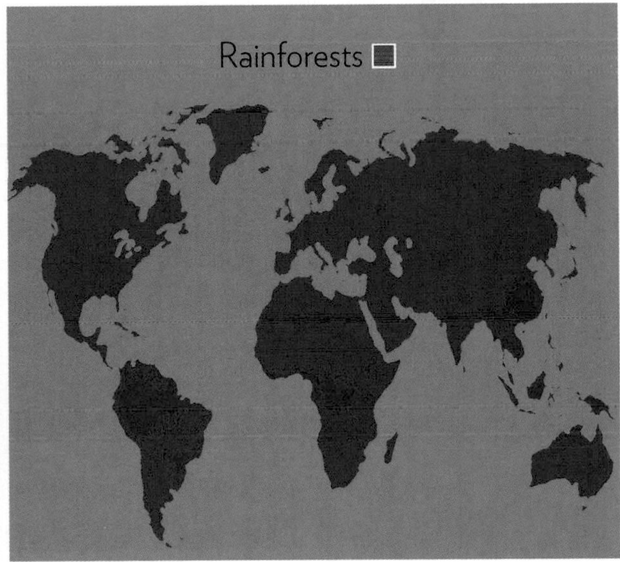

Q:

Could we ever control the weather?

Our civilization remains at the mercy of weather. Our biggest cities and most impressive engineering projects could be wiped out in a matter of hours by an extreme storm. But would controlling the weather make things better—or even worse?

A:

We already have techniques to stimulate clouds to produce rain. But since we don't yet fully understand how weather works, trying to control it might be worse than foolhardy ....

While environmental groups advocate humans altering our behavior and lifestyles to reduce our impact on the planet, and climate-change deniers poke their fingers in their ears and insist there's no problem at all, there is a third group.

These people—many of them respected scientists and engineers—believe humans have the potential to engineer and control the environment, including the weather. They speak of grand plans to bring life to deserts, control rainfall, and manipulate the atmosphere to cool the world.

This process, called geo-engineering, sounds great when it's a plot point in a science fiction novel; but implementing such plans in the real world is fraught with incredible danger.

The weather on Earth is ultimately based on a fairly simple physics formula: force equals mass times acceleration. The severity of weather can be predicted by starting with how much air and water are being forced into a specific area ... and then by adding about a billion secondary equations.

Humans and our supercomputers can already do a reasonable job of forecasting weather up to five days in advance via complex models. We can also make seasonal predictions by looking at long-term trends, such as the cooling and warming Pacific systems called El Niño and La Niña.

But there's much about the weather we still don't understand. We don't know exactly how or why clouds form, especially some of the more complex cloud structures. We don't fully understand lightning. We don't know why some storm systems intensify into hurricanes while others don't. We can guess when tornados might form, but we can't pinpoint where they will touch down. And while we can forecast the probability of rain with reasonable accuracy, we can't tell exactly where showers will fall.

We do know that unsually high rainfall over one region might cause a drought over another. Cool weather in the north can lead to hotter summers in the south. The weather is a single system: changing one part could have unexpected and possibly disastrous consequences for another.

With this in mind, the idea of messing with the weather any time soon seems fairly crazy. But that doesn't stop some people! One of the most widespread weather control techniques is to fire silver iodide or even plain table salt into clouds to make water vapor condense and hopefully fall as rain. This is called cloud seeding. Does it work? It's hard to say for sure, because how do we know whether the clouds would have produced rain anyway? Some studies suggest seeding increases precipitation—both rain and snow—by about 10 percent.

Cloud seeding was used before the Beijing Olympics in 2008, in an attempt to "use up" the clouds and make sure it wouldn't rain on the opening ceremony. Some snowfields use cloud seeding in the hope of increasing snow cover for skiers in peak holiday season.

There are also plans to use seeding, or more exotic ideas like firing jet engines into the sky or dumping liquid nitrogen into the sea, in an attempt to weaken hurricanes.

The problem, though, is that physics equation: force equals mass times acceleration. We want to change the force of the weather, but the sheer amount of mass and acceleration in even a modest-sized thunderstorm dwarfs human capabilities.

That said, the issue of weather modification has rung enough alarm bells to lead the United Nations to ban its use in warfare. Many countries also have laws against weather modification. Could we do it one day? Probably. Should we? Probably not.

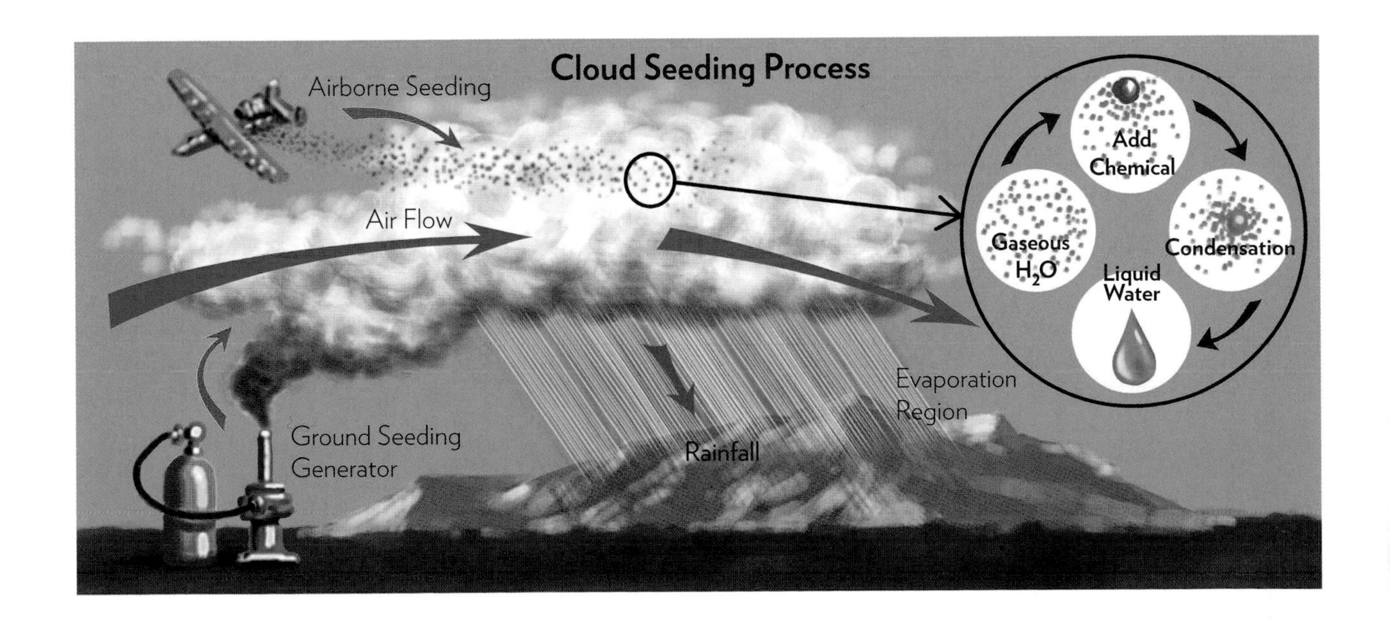

Q: Has anyone ever drilled all the way through Earth's crust?

Earth is mostly a huge ball of molten rock, covered in a thin and fragile solid crust on which we live. To examine the interior of our world, we need only drill through that crust. But it's not exactly simple ....

A:

The Kola Superdeep Borehole, drilled by the Soviet Union, reached a depth of 40,230 feet (12km). Despite this incredible depth, the bore reached only one third of the way through the crust. But even at that depth, things got very, very strange ....

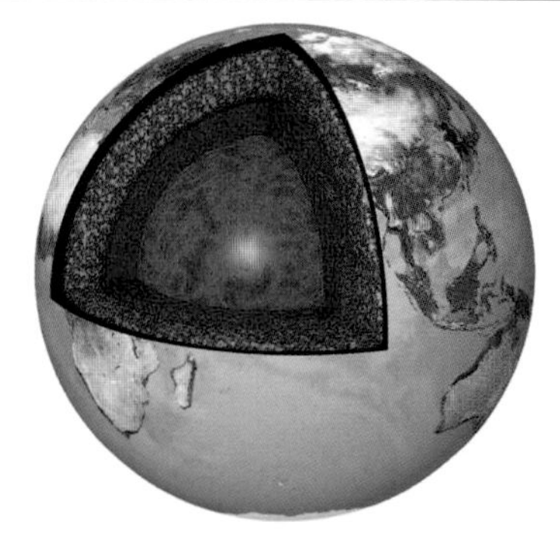

Let's start this answer with some numbers. The Earth has a diameter of 7,917.5 miles (12,742km). The crust varies in thickness from about 3 to 6 miles (5 to 10km) on the seafloor, to 20 to 30 miles (32 to 48km) thick under the continents. In other words, compared to the planet as a whole, the crust is very thin indeed.

Most of the Earth is made of a solid but hot and malleable shell about 1,800 miles (2,900km) thick, called the mantle. It takes up about 84 percent of the Earth's volume. The core of the Earth is made of iron and nickel and has two layers: a liquid outer layer, and a solid inner layer. The core makes up 15 percent of the planet.

That means the crust is just 1 percent of the total mass of the Earth. But humans still struggle to penetrate it to any significant depth.

The closest we've come is the Kola Superdeep Borehole. This drilling project in the former Soviet Union, on the Kola Peninsula east of Norway, managed to get 40,230 feet (12km) into the continental crust.

The effort was immense. Nineteen years of drilling, multiple drill bits, endless engineering challenges, broken drills, secondary shafts ... and at the end of it all the project had made it barely one third of the way through the crust.

Part of the problem was intense heat. Scientists had predicted the crust would be as hot as 212°F. But in fact, the rock was 356°F, and only getting hotter. After reexamining the numbers, the drill team realized that if they were to reach their target depth of 49,000 feet (15km), it would mean working at 570°F. Unfortunately, at that temperature, the drill bit itself would no longer work.

After the fall of the Soviet Union, the Kola Superdeep Borehole project was first mothballed and then abandoned—but not before many fascinating discoveries were made. Not least among these was the discovery that the rock at this extreme depth was absolutely saturated with water. Not from the surface—this water was created millions or even billions of years ago, deep underground, and had remained there trapped by layers of rock.

There was also a huge amount of hydrogen gas released through the shaft of the bore, emitted from the rocks deep in the crust.

Other drilling projects have probed the crust, some with the aim of punching through to the mantle to examine the structure of Earth's interior. Some projects start on the seafloor, so there's less crust to dig through. And a new proposal would see a heat-generating probe literally melt its way through the crust to reach the mantle.

In fact, it's not necessary to drill through the crust if you want to sample the mantle. There are places on the surface where the mantle is exposed, such as in the middle of the Atlantic Ocean.

Mostly, though, we use a combination of seismographs and computer simulations to develop theories about the internal structure of the planet. The way seismic waves reflect off the interior of the Earth gives scientists many clues as to how the mantle and core interact.

In other words, it might not even be necessary to drill through the whole crust to build up a detailed understanding of what lies beneath our feet.

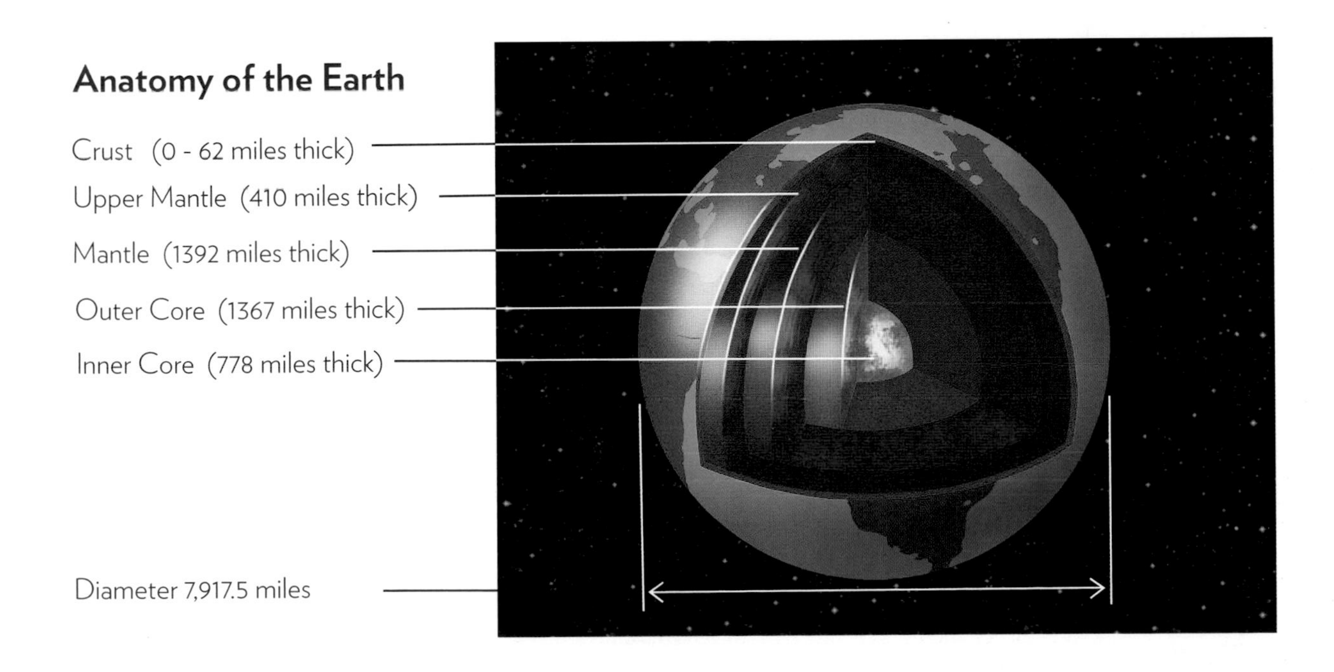

Q: Could we one day travel to the center of the Earth?

A journey to the center of the Earth is a favorite theme in old-school sci-fi, but could we really go there? And what would we find if we did?

A:

The real challenge of going deep isn't heat from molten rock, it's pressure. If we could design a vehicle capable of withstanding unthinkable pressures, then all we need is a big drill ....

Humans have been obsessed with the center of the Earth ever since we got our heads around the fact we live on a big sphere hurtling through space. Some of the theories are pretty out there: in the nineteenth century there were clubs you could join who believed fervently in a sort of mirror-world on the inner surface of the crust. This world had mountains, lakes, seas, weather, its own little Sun, and of course life. You were supposed to get in via a hole at the North or South Pole.

We know now the Earth is a 7,900-mile-wide (12,700km) ball of mostly iron and oxygen with a bunch of other elements tossed in. It has a solid iron-nickel inner core, a liquid iron outer core, and a thick mantle of rock that's in a state geologists call "plastic." That means the mantle is technically solid, but it's gooey and sticky and the rock flows almost like a liquid, causing rising and falling currents—or convection—over thousands of years. On top of the mantle is a thin brittle crust. On top of the crust: us.

Because volcanoes spew red-hot liquid lava, it's easy to imagine the whole mantle must be a vast seething ocean of magma, 1,800 miles (2,896km) deep. But it's actually solid, and something we could—in theory—drill through.

It's very hot—many hundreds of degrees. And it gets hotter the deeper you go. Eventually, in the center, the core is more or less the same temperature as the surface of the Sun—about 9800°F.

To travel to the center of the Earth, we'd need a vehicle capable of withstanding those high temperatures. But the real obstacle would actually be pressure.

Engineers who build submersibles to travel to the deepest parts of the ocean know their vessels must withstand pressures several hundred times greater than at the surface.

At the bottom of Earth's mantle, on the boundary of the outer core where the solid mantle gives way to liquid metal, the pressure is ... a lot—about 136 gigapascals. That's roughly 1.4 *million* times the air pressure at the beach on a summer's day. An unshielded person would be instantly crushed into a thin paste ... which would probably break down into individual atoms because of the heat.

There's lots of good stuff down in the mantle, though. Because of the way the Earth formed and then cooled, many heavy elements sank down into the planet. Some scientists believe there's enough gold and other precious metals in the mantle and core to cover the entire surface of Earth to a depth of 1' foot (46cm).

Geologists have developed their models of the interior structure of Earth by literally listening to the way powerful waves from earthquakes bounce off the various layers of our planet's interior. The way some waves are absorbed, others are bent, and others are reflected allows seismologists to develop models and theories about our planet's true inner self.

As the models become more sophisticated, scientists can match their predictions with actual observations of how the continents move around on the surface, how new crust is made deep in the Pacific and Atlantic oceans, and how earthquakes happen.

Actually traveling to the center of the Earth might be something that never gets out of science-fiction novels. One thing's for sure, though: it would be a hell of a ride.

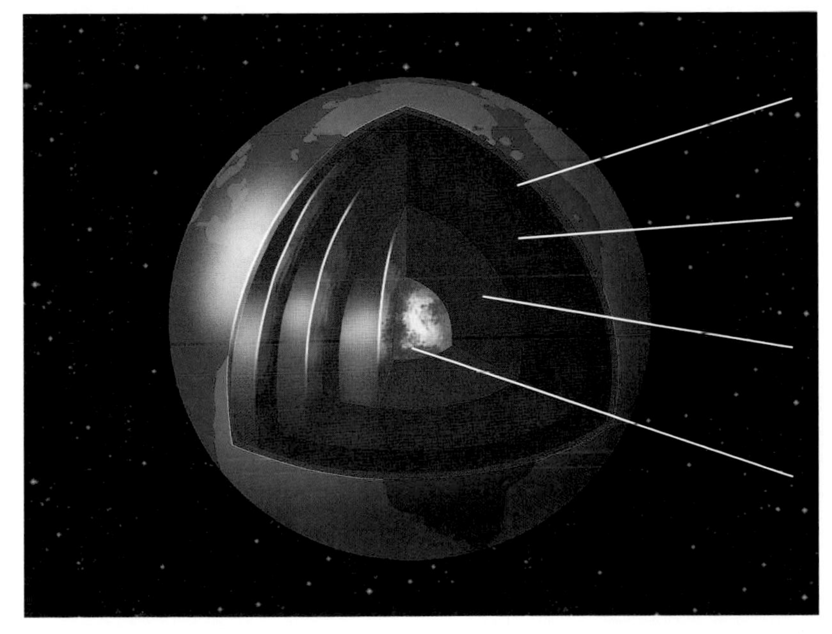

Upper Mantle
(pressure is 24 gigapascals at 2,912°F)

Lower Mantle
(pressure is 136 gigapascals at 6,692°F)

Liquid Core
(pressure is 329 gigapascals at 9,032°F)

Solid Core
(pressure is 364 gigapascals at 9,032°F)

Q:

Why does a compass work?

When you take a long magnet and suspend it on water or on a shaft, it rotates to point magnetic north. This phenomenon has in many ways built our modern world, enabling navigation across long distances. But why does it work?

A:

The Earth's powerful magnetic field is unique among the rocky planets of the inner Solar System. But it doesn't just help us get from A to B, it protects us from many unseen cosmic dangers.

Magnets on Earth, if they're light enough, will spontaneously rotate to point toward the Earth's magnetic north pole because the Earth itself is a giant magnet.

Our liquid metallic core spins at a slightly different rate than the rest of the planet, making the interior of our world a giant electric generator, or dynamo. Convection in the mantle—the huge layer of rock between the crust and core—also adds to this effect. And when a dynamo generates electricity, it also generates a magnetic field.

In addition to being all around us, Earth's magnetic field extends into space, many times the diameter of the planet. If we could see magnetism, Earth would look more like a comet, with a huge tail of electromagnetic force streaming from it.

This magnetic field has proven to be extremely useful to humans and many animals. When we suspend a magnet so it can rotate freely—techniques include floating it in water or attaching it to a pivot as in a compass—the magnet will spin and align facing the north magnetic pole of Earth.

But here's a confusing fact: because the *north* poles of magnets point toward north, and because in magnetism opposite poles attract, in terms of physics and magnetism the "top" of the Earth is actually a *south* magnetic pole! However, to prevent confusion, we refer to it as the north magnetic pole on maps. This confusion came about because humans defined "north" on our maps before we developed a full understanding of magnetism. And of course, because there's no up or down in space, we're free to define whichever end of the planet we like as the top.

Because Earth isn't a perfect sphere (it bulges around the equator), and because we're tilted at a 23-degree angle with respect to our orbit around the Sun, and because the interior of the Earth isn't uniform but has lumpy parts, all this means the *magnetic* north pole isn't at the same place as the physical north pole. The physical North Pole is the point around which the Earth rotates.

What's more, the magnetic north pole moves around—quite a lot. In 2001, scientists pinpointed the magnetic pole near Ellesmore Island in northern Canada. It has since moved beyond Canada toward Russia, at a speed of about 35 miles (56km) a year.

You can tell when you're standing on the magnetic north pole because if you hold your compass out, the needle will try to point straight *down* into the ground.

The magnetic field of Earth—scientists call it the magnetosphere—does more for us than let us figure out which way is north.

The lines of magnetic force that flow out into space around us actually prevent certain kinds of particles from reaching the surface of the planet. The Sun, apart from providing light and heat, also blasts Earth with dangerous radiation. It's the magnetosphere that protects us from the more dangerous particles and stops them from stripping our atmosphere. The magnetic shield is so effective, spacecraft designers are thinking of ways to have a spacecraft generate its own mini magnetic field to act as a radiation shield for long journeys, such as to Mars.

Speaking of Mars, the Red Planet has no significant magnetic field, and scientists believe that's why it no longer has an atmosphere or oceans—these have been blasted away by radiation over millions of years.

The north magnetic pole is actually the SOUTH pole of the Earth's magnetic field

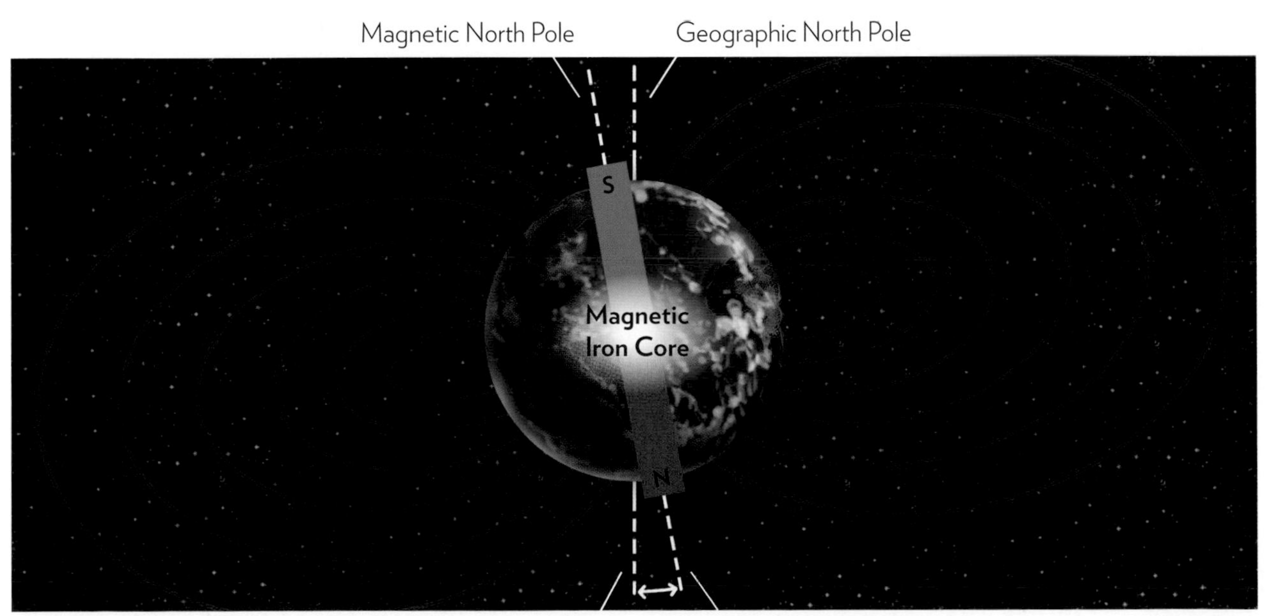

Could the north and south poles really switch?

The magnetic poles move around the surface of the Earth. Could they ever completely switch, so that north became south and south became north?

A:

The magnetic poles do switch, and if you're talking geological timescales, they switch fairly regularly. Figuring out when the next switch will occur, though, might be impossible ....

The Earth's magnetic field randomly changes direction over timescales of a few tens of thousands to millions of years. Each period is called a "chron," with each flip of the direction of the magnetic field being the start of a new chron.

Our magnetic field, which protects us from harmful radiation coming from the Sun and other objects in space, has two distinct poles: north and south.

Lines of magnetic force flow between the poles, and as a result any magnetic material on the surface tends to align itself with the magnetic field. This is why our compasses work: the magnet inside turns to point along the lines of force (see previous pages for more).

All magnets have a north and south pole, and if two magnets are close together, their opposite poles will attract and—if the magnets are strong enough—stick together. So one magnet's north pole will attach to another magnet's south pole.

If the polarity of one magnet is reversed (most easily by just turning it around!) so that the south poles are facing each other, the magnets will move apart.

When one magnet is much bigger than the other (as in the case with a tiny compass needle and the entirety of our planet), then the smaller compass just turns to align itself in the direction of the bigger magnet's opposite pole.

When the Earth's magnetic field reverses, compass needles will swing around and point south. This won't be a big problem, as people will simply adjust the labeling on their compasses and continue as normal. The compass is still pointing reliably in a single direction, which is what enables navigation (see the next page for more on this).

Unfortunately, when the poles do reverse, they won't necessarily do it instantly. There can be periods when the magnetic field more or less shuts down, to as little as 5 percent of its maximum strength.

Instead of two distinct poles, the magnetic field could have several poles that move around chaotically until a stable field returns.

Computer modeling of the magnetic field shows that a normal north-south (or south-north) field is the most stable, so the Earth's giant magnet usually ends up like this.

What causes these reversals? Because the Earth is not a perfectly geometric structure and has many odd lumps and bumps and different densities and irregularities, there's inherent instability in the way the core generates our magnetic field.

Again, extremely detailed computer modeling of the internal structure of the Earth and our magnetic field actually shows that magnetic field reversals—poles swapping—occur over long enough time periods. The period between reversals is quite random—sometimes every 10 thousand years, sometimes every 10 million.

This is backed up by evidence from rocks on the seabed, which show lines of magnetic alignment from when they were formed at mid-ocean ridges. When molten rock comes up from the mantle, its various magnetic elements are locked into a particular configuration—based on the direction of the magnetic field—as the rock cools.

Rocks from different time periods show different magnetic alignments. Pole swaps are just a normal part of life on Earth.

Q:

If the poles switch, what would happen to our compasses?

Geological evidence shows that magnetic field reversals—the switching of our north and south magnetic poles—are quite normal. But we've never lived through one. Would our civilization be disrupted?

A:

If the poles switch and stabilize in a reversed direction—north is south and south is north—then there's no real problem. We just change the labels on our compasses; they still point in one reliable direction. But if the pole switch takes a long time and the magnetic field loses strength, that could be a worry ....

Quite a lot of the machinery of our society relies on Earth having a strong magnetic field. Outside the safe envelope of the atmosphere and the magnetosphere, the universe is a hostile place.

As the Solar System moves through space orbiting the galactic center, all sorts of nasty high-energy particles come sleeting through. Things like x-rays and gamma rays, which can give us cancer if they hit our fragile bodies.

Fortunately, almost all of these particles are bounced off our magnetic field. The magnetosphere also protects us against the more harmful parts of the Sun's energy output and defends against solar storms and flares.

The good news is that if the magnetic field collapses or drops in intensity during field reversal, modeling shows that the solar wind—a stream of particles constantly flowing from the Sun—will interact with our ionosphere in such a way as to keep up enough of a shield to protect us against outer-space nasties.

Down on the surface, our compasses will no longer point in a specific direction. Some models suggest that during a reversal, the magnetosphere will develop several weaker magnetic poles that may move around on an almost daily basis. This would be very irritating for anyone trying to navigate with a compass, because the needle could be pointing a different direction each time the navigator reaches for a map.

There are other ways of navigating, though. There are techniques using the Sun and the stars that don't need compasses, though they do rely on accurate timekeeping. Still, as long as our clocks still work, we should be able to figure out where we are on the map with some careful observations and a little math.

In fact, today, while all ships are supposed to carry compasses, most of the big transport vessels rely entirely on GPS for navigation (but there's always an enthusiastic officer who knows how to use a sextant!). The loss of our magnetic field won't affect GPS directly, but there's a risk that the satellites could be damaged by radiation, especially from solar storms due to the weakened shield-effect the magnetosphere currently provides.

Any magnetic field chaos could last for many hundreds of years, if the current understanding about reversals is correct. That's enough time for us to respond to the challenge and launch, say, a new fleet of radiation-shielded navigation satellites.

Because magnetic field reversals seem to match up with some of the big extinction events in the past, scientists have worried that losing our magnetic field could spell doom for many species.

But there are other times when lots of reversals have occurred—as many as 50 in a period of just a few million years—and there's no corresponding mass extinction. At the moment, it looks like those reversal/extinction match-ups are just coincidence. Or, as seems likely, increased volcanic activity, which causes extinctions, could also cause a magnetic field reversal.

Should you worry? Probably not. A reversal could happen tomorrow, or it might not happen for another 10 million years. And unless you're a navigator or a compass salesman, such a reversal may not even affect you.

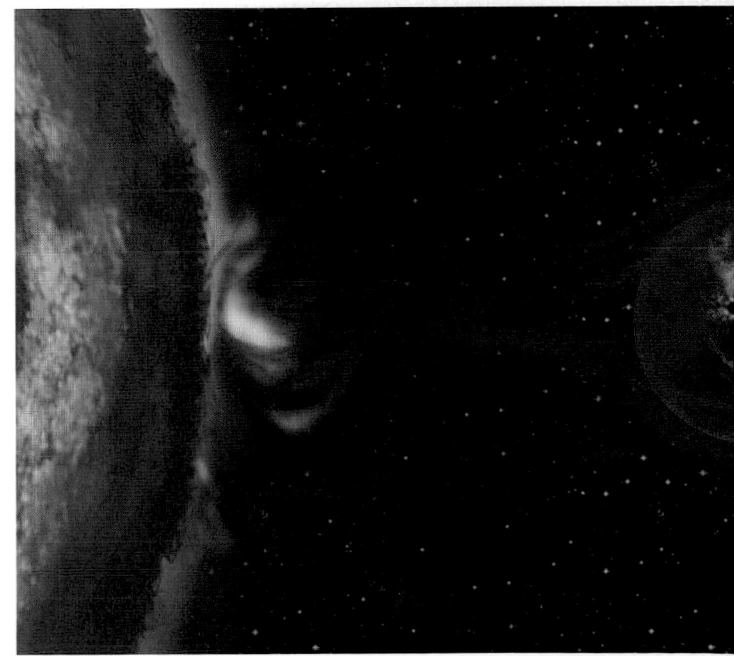

Q:

Does a single large volcanic eruption pollute the atmosphere much more than all human industry ever has?

Along with ash and lava, volcanoes release millions of gallons of toxic gasses, including carbon dioxide and sulfur dioxide. But how does this natural pollution stack up against human sources of toxins?

A:

The pollution caused by any of the volcanic eruptions of the last couple hundred years is less than the pollution put out by human industry. Super-eruptions like Krakatoa, Toba, and Yellowstone, though, are another story ....

Opponents of the theory of man-made climate change often point to volcanoes as a much larger source of greenhouse gas emissions than human industry. Leaving aside the question of why this would make it okay to pollute the planet anyway, the numbers from volcanoes are less impressive than people might think.

Part of this has to do with the sheer scale of human industry and transportation. With a global population of seven billion and with the vast majority of countries having at least some kind of CO_2-emitting industry, the human output of CO_2 per year is somewhere in the order of 35 billion metric tons and climbing.

Volcanoes, on the other hand, put out only 200 million tons of CO_2. So yes, a good-size volcano can rival the CO_2 output of a single large company or maybe even a moderate-sized city, but once you start adding in our transportation and agriculture, they quickly fall behind.

On the subject of climate change, volcanic activity also releases sulfur dioxide, which actually blocks sunlight and reduces global warming. It can also turn into acid rain, though, which is less helpful.

A big eruption will also throw dust and ash high into the atmosphere, creating a temporary shield that will dramatically reduce the amount of sunlight reaching the surface. The result of a serious eruption isn't global warming—it's the exact opposite. Temperatures will drop, but only until the dust, ash, and sulfur dioxide fall back to the surface.

That figure of 200 million metric tons of CO_2 includes all volcanoes on land and also the many deep-sea volcanoes in places like the Pacific Ocean's famous Ring of Fire.

There are currently about 70 active volcanoes in the world. For volcanic activity to beat human CO_2 output, the planet would need an extra 10,000 or more spewing gas and lava.

In 2010, an Icelandic volcano with the tongue-twisting name Eyjafjallajökull erupted. It put enough ash, dust, and even tiny particles of molten glass into the air to shut down most of Europe's air travel. Naturally, the volcano also pumped many hundreds of thousands of tons of CO_2 into the atmosphere.

While estimates of the precise volume of CO_2 from the Icelandic volcano are approximate, scientists believe that when you add the volcanic emissions but then subtract the amount of CO_2 that wasn't emitted from the jet engines not flying during the eruption, the planet actually ended up with *less* CO_2 in the atmosphere than if the volcano hadn't erupted.

That's not to say that volcanism doesn't have the potential to dwarf human industrial output of CO_2. Supervolcanoes like the Yellowstone caldera and vast regions of molten rock called "large igneous provinces" have in the past released millions upon millions of tons of carbon dioxide and other chemicals, perhaps in just a few months and radically changed the makeup of the atmosphere.

These mega-eruptions pump thousands of cubic *miles* of dust and ash into the sky and shroud the planet in a blanket of grey. This blocks the Sun for years and plunges us into a snap-freeze. The ash eventually falls, smothering the land and killing anything that survived the cold. It's possible that volcanic activity in Siberia was at least partly responsible for one of the biggest extinction events ever—the Permian Extinction—which killed 95 percent of land-based life.

Volcanoes might not play much of a role in global warming, but they could still kill us all.

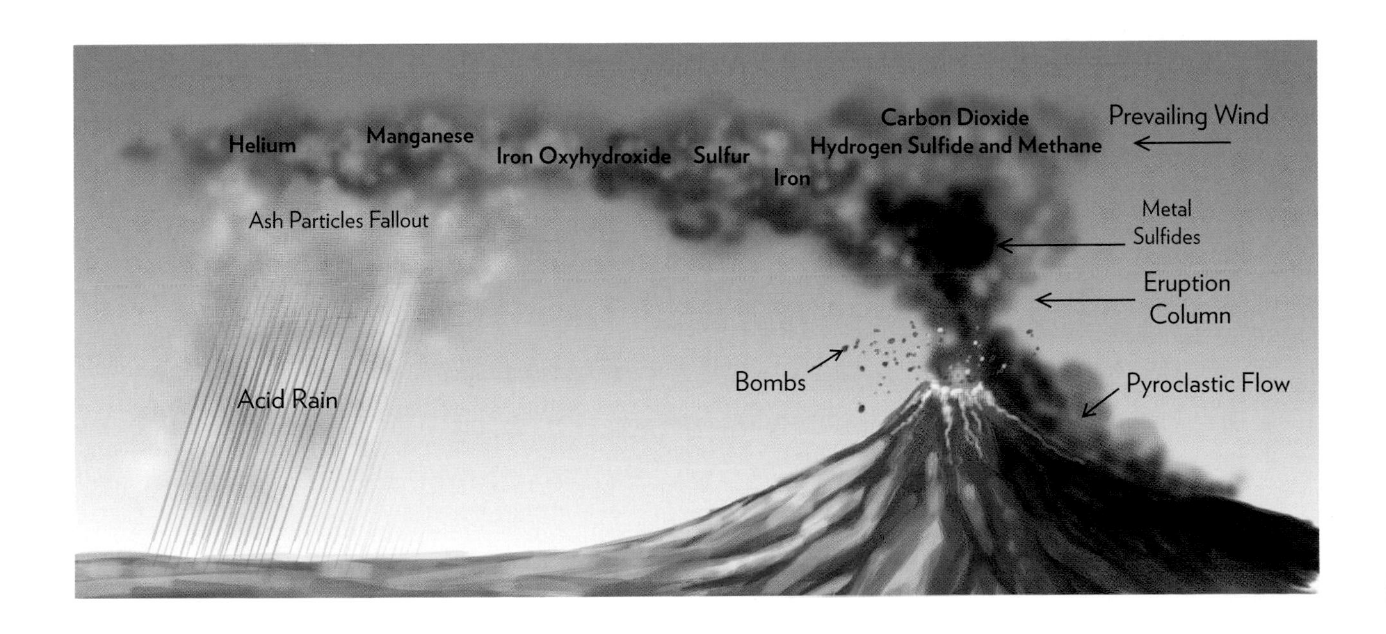

Q:

If we reverse climate change, could the sea level drop dramatically?

As the Earth warms, glaciers and ice caps melt, adding their water to the ocean. Plus, a hotter sea expands, further raising sea levels. If we master the art of reducing global warming, could we end up with a much lower ocean and a whole new set of problems?

A:

The sea level on Earth changes over time, swinging from extreme highs to extreme lows. The changes humans are making are minimal, but they could still be disastrous for us ....

Over geological timescales—millions of years—there's nothing static about the surface of the Earth. Continents move around. Mountain ranges are pushed into the sky and eroded back down by rain and wind. The ocean itself ebbs and flows, rushing to cover huge areas of land and then retreating many miles from previous coastlines.

That's big-picture, deep-time stuff. But on smaller scales—less than a million years—there's still plenty of change. One of the biggest variables is the sea level.

Earth routinely moves in and out of so-called ice ages. When global temperatures drop by several degrees, more ice forms in high latitudes. This ice is drawn, via evaporation, from the ocean. More ice on land, less water in the sea, so sea levels drop.

We're currently living in a post-glacial world. A geological spring, if you like, of a planet recovering from an ice age of pretty average intensity. In fact, it's likely the planet would be either stable or warming slightly even without human input. But as the ice melts, the water returns to the sea and the sea rises. Water also expands as it warms, and in an ocean, even a few degrees is enough to raise the surface by several feet.

Only 8,500 years ago, there was a broad sweep of land between England and the west coast of the Netherlands. Archaeologists call it Doggerland after the Dogger Bank, which is now a fishing ground.

Seabed archaeological digs have found lots of stone tools in Doggerland, along with the remains of animals like deer and lion (and human, too). In fact, most of the really good archaeological sites for stone-age human remains are actually underwater, just off the coasts of Europe and far eastern Russia.

The point here is that before humans even developed the technology to start pumping CO_2 into the air, we survived a catastrophic sea level rise. Some scientists estimate we lost 40 percent of our hunting grounds to the rising tide. The land bridge from Russia to Alaska was flooded, along with a huge plain between Papua New Guinea and Australia. The sea may have risen as much as 300 feet in the last 10,000 years as the last of the ice sheets melted.

Today, we live on the edge of a drowned landscape. Even if we cease CO_2 production and return the atmosphere to the precise state it was in back in, say, 1800, it's unlikely the sea would drop significantly.

Reclaiming those ancient flooded countries would mean the planet would have to go back into an ice age. Yes, we'd get back Doggerland, but we'd lose all of Canada—and the United States to below Chicago—under ice sheets.

If the sea rises and falls naturally, why are we so worried about human-induced sea level change? Because we've built so much infrastructure so close to the coast. Only a couple dozen feet of extra depth—barely a statistical glitch on the scale of the planet's entire history—could leave New York flooded and do trillions of dollars' worth of damage.

It's likely that one day, hopefully thousands of years from now, we will have to face the challenge of really significant sea level change. We should consider reducing man-made increase as a practice now.

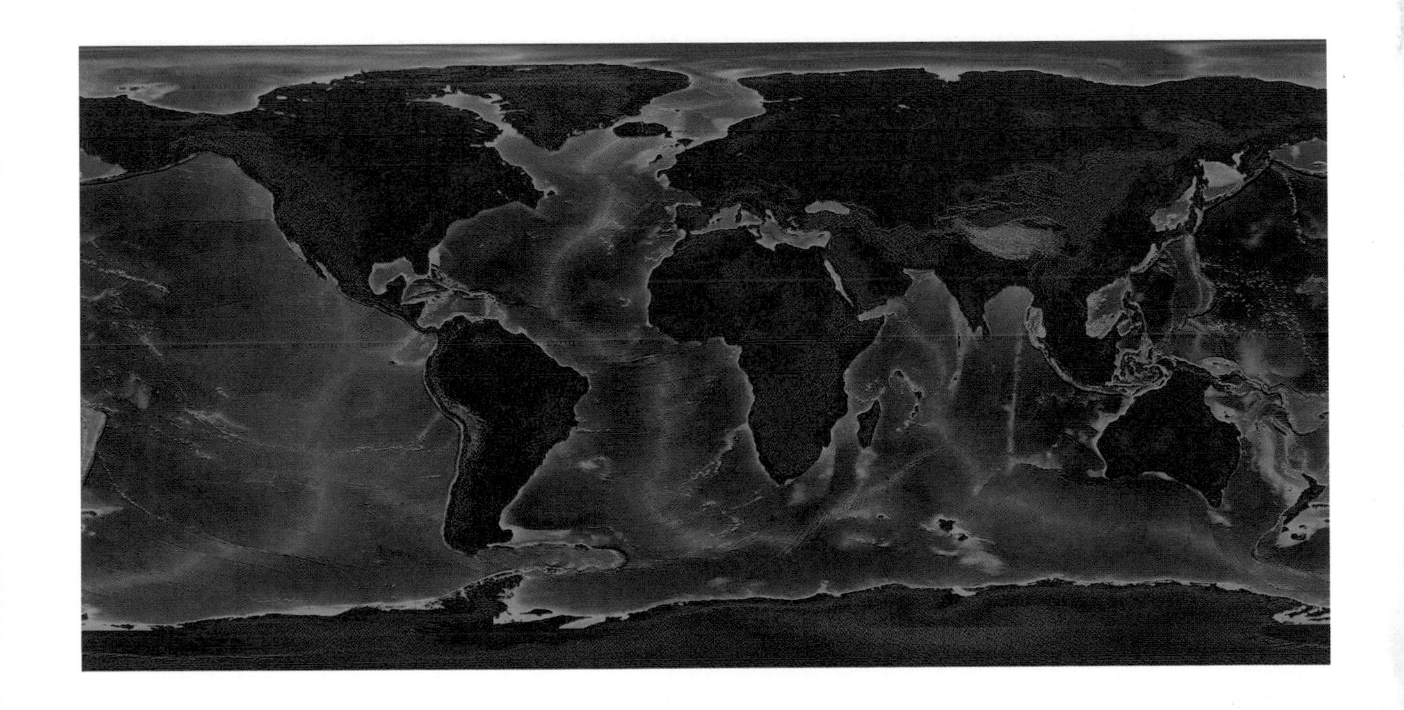

Q:

Could the ocean ever freeze completely solid?

Earth has a liquid ocean because our orbit is just the right distance from the Sun. But what if we wandered farther out, or something blocked the sunlight? Could the ocean freeze totally solid?

A:

Without the Sun, ice would form on the ocean to great depths. But since water freezes from the top down, and the Earth produces heat from its interior, even without the Sun we might hold on to some liquid water ….

One of the fascinating properties of water is that it freezes from the top down. This happens because water has the ability to become "supercooled." It also becomes less dense as it cools below 39°F.

This means that very cold water floats to the top, where it forms ice. This layer of supercooled water then insulates the slightly warmer water below it, delaying freezing. Ice slowly crystallizes its way to the bottom of the water column until the volume of water is entirely frozen solid.

When the water is salty, it's even harder to freeze—for a start, saltwater has a lower freezing point. As saltwater freezes, the salt is excluded from the ice. It mixes with the remaining liquid water, making that water even denser and saltier and further lowering its freezing point. The more you freeze the sea, the harder it gets to freeze.

Scientists are almost positive that Jupiter's moon Europa has a liquid water ocean beneath its icy crust, despite the fact that sunlight there is a mere fraction as strong as it is on Earth. But on Europa, the sea is kept liquid by forces other than the Sun's heat.

As Europe orbits Jupiter, it gets pulled and stressed by the giant planet's gravity. This tidal flexing is enough to produce heat and melt ice. What's more, Europa may have a molten core, and heat from that core could seep into the ocean and keep it fluid.

The same applies to Earth. Our hot mantle, liquid metal outer core, and heat generated by radioactive decay deep underground all contribute to the planet's heat output. Ours is a warm world, even without the Sun shining down on us.

Averaged across the whole crust, the heat from the planet's interior is only 8.7 milliwatts per square foot—less than a tenth of a percent of the heat we get from the Sun.

But this heat is concentrated in areas where the mantle is exposed, such as around undersea volcanoes and the mid-ocean ridges. Here, water can be heated beyond boiling point, and only remains liquid because it's under such huge pressure.

Miles below the surface, there's no sunlight whatsoever; yet life clusters and thrives around vents and so-called "black smokers"—black chimneys that spew superheated water from deep in the crust, rich with minerals.

Even though it would be nearly impossible for the ocean to freeze totally solid with the Earth still producing so much heat and sitting in its nice warm orbit, there have been times in the past when the whole surface has certainly iced over.

Earth has at least three major types of climate—Greenhouse Earth, Icehouse Earth, and Snowball Earth.

Greenhouse Earth is a hot, humid world with lush jungles, high sea levels, and lots of CO_2 in the atmosphere. Icehouse Earth has big ice sheets, low sea levels, big deserts, and less CO_2. Snowball Earth is a white globe completely covered in ice. The last Snowball Earth happened at least 650 million years ago, just before the evolution of multi-cellular life. That's right: the near-freezing of the oceans may have given life the kick in the pants it needed to evolve from microbes into humans.

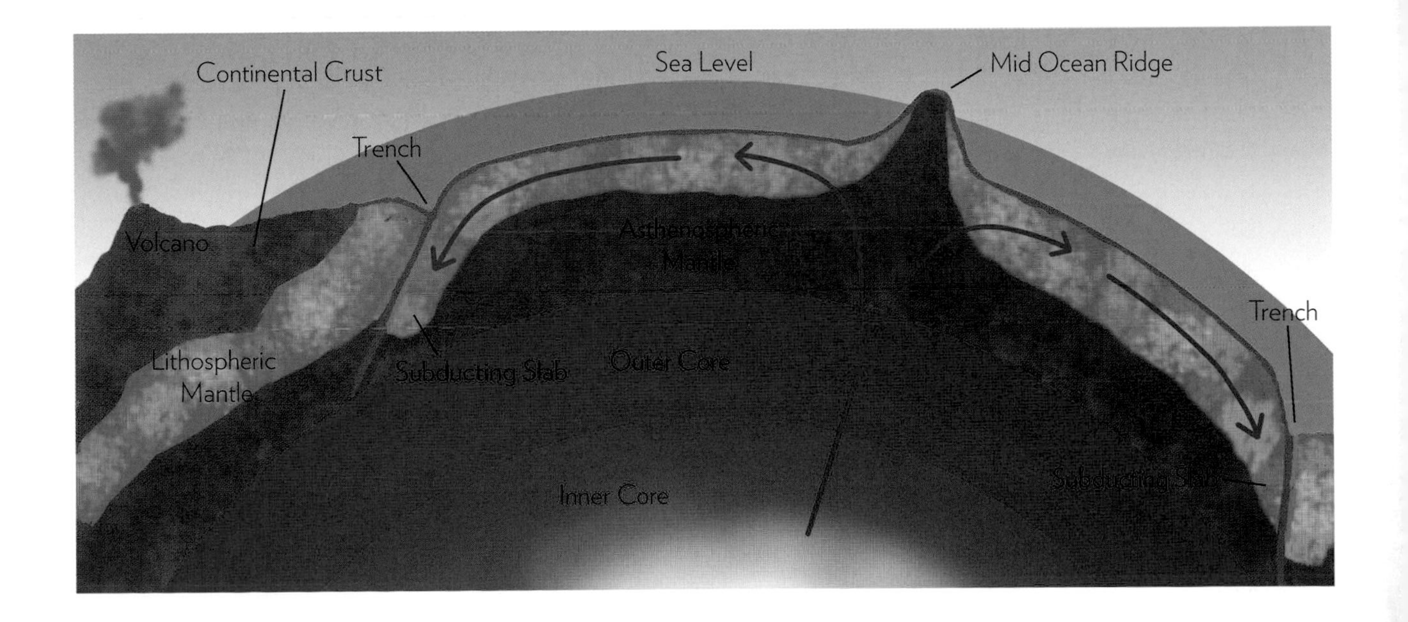

Q: How can we be sure there wasn't a technological civilization living on Earth millions of years ago?

Are humans really the first technology-using animal to walk the Earth? With a billion years of history, it seems pretty unlikely! Though surely if smart city builders had lived here before us, there'd be some kind of sign ....

A:

The fossil record has nothing in it to indicate a technological species came before us. But we've made some very particular changes to the world that should remain for millions of years—a sort of technological fingerprint for later life forms to discover .....

To be totally scientific about this, we have to say that it's still *possible* we are not the first high-tech species to live on Earth. And there is, hidden away somewhere in the geological record, evidence of super-smart dinosaurs or something similar.

After all, humans are just another kind of mammal, and our species may even be less than a million years old. Surely in the 135-million-year history of the dinosaurs, there was at least one species that used tools, made fire, built houses ... no?

To explain why there almost definitely hasn't been another high-tech animal on this planet, it's helpful to look at what humans would leave behind if we all left or died out in the next few centuries.

For anyone visiting in the next hundred thousand years or so, the evidence of human habitation will be pretty plain. Our cities will be buried under plants and our roads long since eroded away, but alien scientists will, with a little digging, be able to uncover all sorts of junk—especially plastics, toxic wastes, and certain metal objects.

After millions of years have passed, our former stewardship of Earth will be more difficult to detect. But good scientists will still be able to spot clues. Our quarries and mines, with their unusual geometric fracturing of hard rock, should endure for millions of years, though they will be filled with sediment. Deep-penetrating radar should be able to detect them, though.

There will also be unusual deposits of pure metals, because we mined ore and refined it into pure elemental metal. The Earth will be strangely lacking in radioactive isotopes of uranium on the surface—we used it in nuclear reactors and weapons. And the distribution of such rare-earth elements as lithium will be odd, too, because we mined it and made it into batteries and other things.

Some of our metal tools, machines, and art could survive for millions of years, especially bronze statues. And if we do die out rather than leave, our legacy will be preserved in the fossil record.

If fossils can show detail as fine as individual feathers and the points where muscle anchored onto bone, it's likely human civilization will leave all sorts of intriguing shapes in rock strata.

Because of all this, it seems reasonable to assume that if there had been a city-dwelling, jet-plane-flying, nuclear-reactor-building, high-tech civilization on Earth before humans, evidence of this kind would remain. We would see their mines, their machines, their culture preserved in the rock, if nothing else. But as far as we can tell, the Earth really was "primordial"—untouched by technology—before humans evolved.

On the other hand, we should never underestimate the erasing power of Earth's tectonic and seismic activity. Much of the rock on the surface is new—geologically speaking—and the signs of a prior civilization could have been recycled back into the mantle by now. The fossil record is, after all, enormously patchy.

For now, though, it looks like humans are indeed the first technological species to roam the Earth. Let's hope we're not the last.

life science

The world is full to bursting with living things, but what makes them tick?

Life is what makes Earth special. No other planet yet discovered has such an amazing diversity or sheer mass of life. No matter where you go, from the coldest ice sheet to the driest desert, you'll find life—though sometimes you'll need to pack a microscope.

What makes something alive? What's the scientific definition of life? We don't even have a very good idea of where to draw the line between living and nonliving, at least on a microscopic scale. Not everything breathes oxygen, not everything ages and dies, not everything reproduces in ways we fully understand.

Even though much about life remains a mystery, our understanding grows by the day. We've discovered such amazing things as DNA, the mechanisms by which we age, how plants are able to make food from sunlight, why some animals are so large, and much more ....

Q:

What is the earliest evidence we have of life on Earth?

Our understanding of when and how life first appeared on Earth continues to improve. Evidence today points to life appearing almost as soon as the Earth's crust solidified enough to support it. But how can we be so sure?

A:

Coral-like structures called stromatolites provide some of the earliest evidence of life and date back 3.5 billion years. But figuring out the age of a stromatolite is anything but straightforward ....

Off the coasts of certain shallow seas and in some lakes, you can find curiously shaped mineral deposits. Not quite coral, not quite rock, nevertheless it seems obvious to look at them that they were made by some kind of life.

Called *stromatolite* (from a Greek word for "bed-like rock") these odd formations are made by microorganisms such as blue-green algae. They range in form from towering cones to round pillow-shaped structures, or uninteresting vaguely rounded collections of tiny grains all cemented together. They're made by tiny, single-celled creatures that put out a mucus, which then picks up grains of silt. As the microorganisms build their calcium carbonate bodies (like modern coral), the silt gets glued into the structure. Over time—lots and lots of time—layers of silt build up into distinctive domes, columns, and other shapes.

These are pretty basic life forms. It's not a sophisticated colony of complex creatures, but rather a biological "mat"—a layer of scum that slowly grows over the remains of the previous layer of scum. Hardly exciting ... unless you're a paleontologist!

Paleontologists use radiometric dating to figure out how old a rock sample is. The problem with fossils is that they're usually made of types of rock that don't contain the necessary radioactive particles for dating.

In this case, scientists compare the fossil to the rocks around it. If the fossil is between two layers of rock that *can* be dated, it seems common sense to assume the fossil is aged somewhere between the two rocks. This is why you often see descriptions of dinosaurs like "This species lived 80 to 95 million years ago."

Stromatolites are extremely common in the fossil record, and they exist at many different layers. They are an excellent constant in the story of evolution.

Around 3.5 billion years ago the stromatolites absolutely dominated the biosphere. Life, it seemed, was all about stromatolites.

By examining how modern stromatolites live and grow, our theory of evolution now suggests they were responsible for producing a lot of the oxygen in our atmosphere. Even today, phytoplankton and cyanobacteria (the scientific name for blue-green algae) pump out billions of gallons of oxygen.

That's right, stromatolites still exist today—you can see living examples in Western Australia, the Bahamas, British Columbia (Canada), and a few other places. But it's their presence in the fossil record that gets scientists excited.

But our theories about the origin of life aren't entirely tied to this one type of ancient microbe. We can look at the DNA of modern life forms and make assumptions about how long it's been since any two species were closely related.

By "reverse evolving" modern life, scientists can see that all life on the planet had a common ancestor that, given the apparent rate of evolutionary change, must have lived at least 3.5 billion years ago.

What's interesting about this number is it suggests Earth is such a perfect place for life to grow that living things appeared as early as possible—only a billion years or so after the planet formed. The crust may not even have been entirely solid. Seas of lava could have jostled for space with warm, shallow seas. And these seas were already full of chemicals just itching to combine and eventually form the amazing biodiversity we see today.

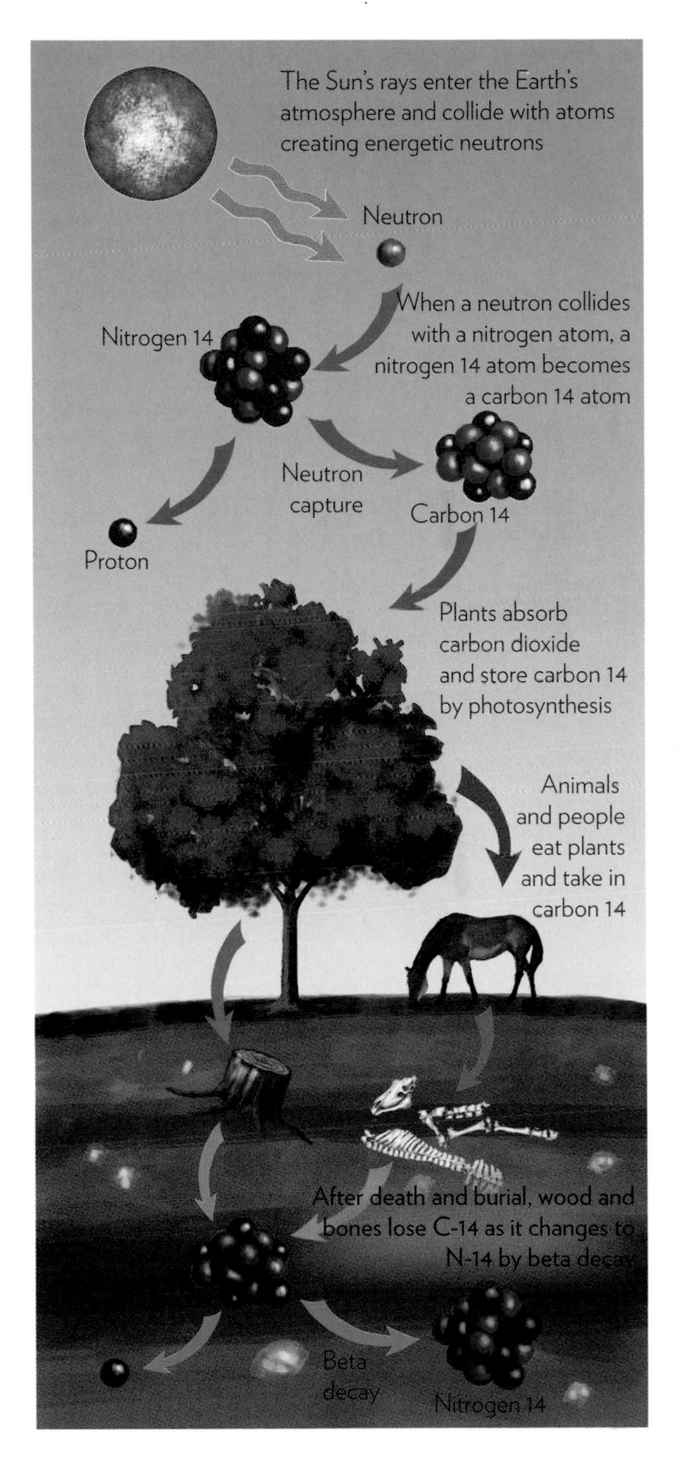

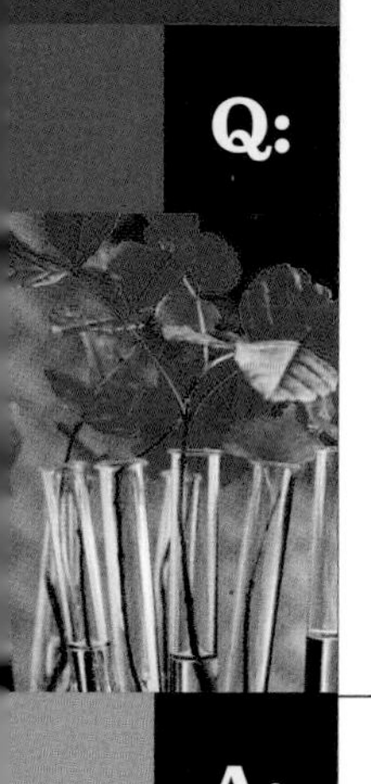

Q:

Why does every living thing need water to survive?

What does every organism on Earth have in common? The need to breathe oxygen? Nope. Access to either the Sun itself or to something that grew from the Sun? Nope. The answer is water. Why is water the key to life?

A:

Water has unique chemical and physical properties that make it the perfect medium in which to mix other chemicals, transport energy, remove waste, and a whole bunch of other useful things.

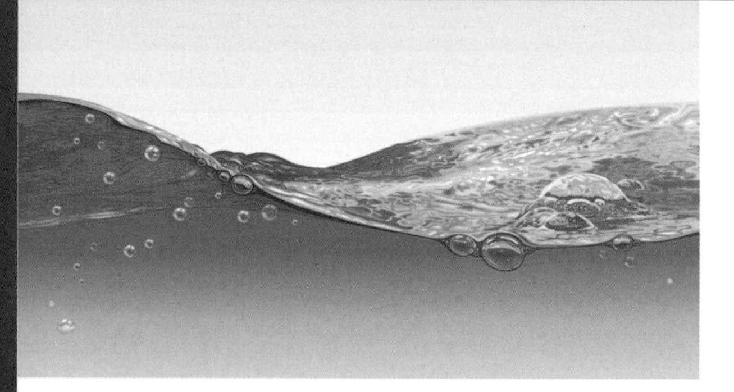

Take one oxygen atom and add two hydrogen atoms and what do you get? An entire biosphere, millions of different species, and a planet full to overflowing with life.

Water is the key ingredient to life and is used by every single living thing ever discovered. Even the toughest "extremophile" bacteria like the ones that eat concrete and excrete sulfuric acid still need water to go about their lives.

Water is a solvent, a liquid that allows other chemicals to mix into it without actually reacting with those chemicals and changing them into something else (though life certainly does use water as a *reactant,* too, a sort of engine room for chemical reactions). You can dissolve oxygen in water, which makes it ideal for transporting the otherwise explosive gas into living cells.

Indeed, water is the basis of blood—around 83 percent of it, actually—which can carry energy and building materials through the body of a large organism like a human. And you can use water to reduce the concentration of a chemical and flush it out of an organism's body in the form of urine.

On the microscopic scale, the movement of water in and out of cells is fundamental to a living thing growing and moving. Water full of dissolved molecules can be pumped into a cell, the molecules removed, unwanted molecules added, and then pumped back out again.

Life formed in the first place when different organic compounds mixed together in the ocean. As compounds bumped into each other, they stuck and reacted. Eventually they became large and complex enough to start reproducing—but this was basically just a whole bunch of chemical reactions. And many of those reactions only work in the presence of water.

For large-scale life to exist (life that isn't microscopic—everything from fleas to elephants), it needs to be able to get lots of energy for chemical reactions. Transporting energy through solid matter is too slow. Through a gas is too chaotic and difficult to control. Some liquids won't hold on to energy; others boil or freeze too easily or change properties too much at different temperatures.

Water does none of this. It can carry energy and conduct electricity—though not too much electricity, which would be bad, too. It can be acidic, or it can be alkaline and thus play a role in a massive number of chemical reactions. In short, it gives life the flexibility to be all it can be. And here on Earth you can find liquid water anywhere, even at the frozen poles.

Could life exist without water? There are other liquids that might do the same job of transporting materials and removing waste. Ammonia is a candidate, but it's only a liquid at very cold temperatures (it boils at -28°F) so any ammonia-based life would move very slowly compared to us.

Some of the hydrocarbons (chemicals with hydrogen and carbon in them) could also work as solvents, and if there is life on Saturn's moon Titan as some have suggested, it could use hydrocarbons like ammonia in its cells. There's a puzzling lack of hydrogen in Titan's lower atmosphere, which might be evidence of life "breathing" it to react with hydrocarbons.

Here on Earth, water remains the key to life. We evolved in the sea, and today we carry trillions of tiny oceans around with us—one in each water-filled cell.

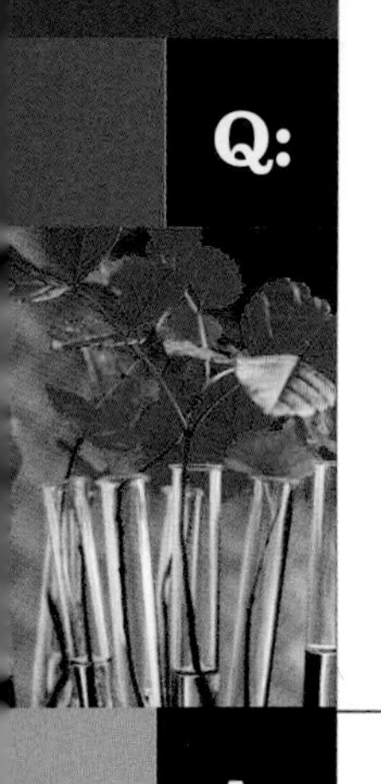

Q:

Why isn't DNA perfect? Why are there mutations?

Plants and animals are able to grow thanks to DNA code that gives their cells instructions on how to build tissues, nerves, and other internal structures. But DNA doesn't always work properly—errors occur in the replication process. These are called mutations, but how and why do they occur?

A:

DNA relies on complex chemical reactions to copy itself, and because there are so many atoms involved, the process isn't 100 percent accurate. But DNA has an amazing ability to correct most of the errors that happen ....

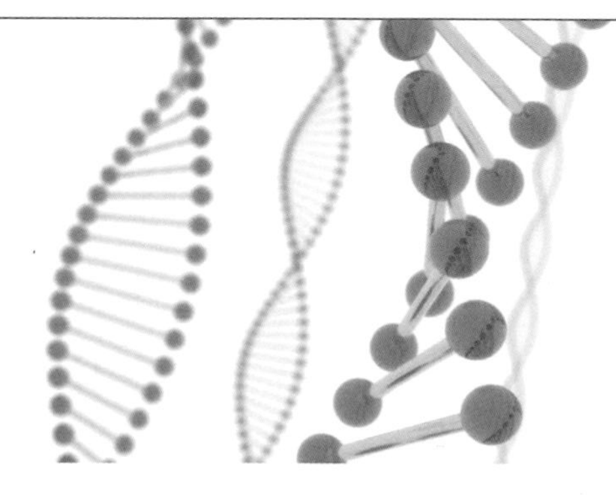

The way DNA provides instructions for an organism's growth and reproduction, copies itself, and combines with other DNA to make new plants and animals with a unique genetic code is one of the most amazing aspects of biology.

DNA is essentially a molecular code made up of around 440 million sets of instructions. It forms a distinct "double helix" spiral shape, like a ladder twisted around itself. Each rung of the ladder provides vital information for the growth of the organism to which the DNA belongs.

Plants and animals grow by dividing and replicating their cells. Life that reproduces sexually starts with just two cells—an egg and a sperm—but by the time the life form is mature, it will consist of trillions of cells. Each one has been assembled according to the instructions in the life form's DNA.

Humans are *eukaryotes*, because our cells have a nucleus in the middle. Eukaryote means "good nut" or "good kernel" in Greek. This nucleus contains our genetic material, which is a mix of DNA and other molecules that, when combined, are called *chromosomes*.

When your body needs to replace material (we replace many—but not all—of our cells about every 10 years), cells will grow and split into two new cells in a process called *mitosis*.

At one stage of cell division, the DNA in the nucleus makes a copy of itself for the new cell. Even though DNA is a molecule, it's a huge one, consisting of more than 15 billion atoms. So it can be forgiven for not making an exact copy every time!

However, because creating a new cell depends on the copy being correct, DNA can actually "proofread" itself. If it detects mismatches in the new strand, it will undo the work and redo it. It's an amazing process that makes complex life possible.

But the proofreading system itself isn't perfect, either, and at the end of cell division, minor errors can creep through. Many of these errors have no effect, but some can be significant enough to change the way a new cell forms. These are called mutations.

In the vast majority of cases, a mutation will kill the new cell. The body flushes the cell and tries again. Total amount of time, material, and energy wasted? Very little.

Sometimes, though, the mutation doesn't kill the new cell. The worst-case scenario is a new kind of cell that divides and divides again, out of control in the body, impacting on important tissues and organs. This is cancer.

A more significant kind of mutation is one that occurs in a sex cell like a sperm or an egg. These cells carry only half the number of chromosomes as a normal cell, because they will build a new set of chromosomes when the egg is joined with a sperm. This kind of mutation can be passed on to the embryo.

Mutations are not inherently good or bad. Some of them will make a child sick, as in the genes that cause cystic fibrosis or type 1 diabetes. Others are beneficial, like the European mutation that gives people lactose tolerance and the ability to drink cow's milk. Or they might just change the way we look, like giving us blue eyes.

Over very long time periods, mutations build up, and the organism changes and becomes a new species. This is the essence of evolution.

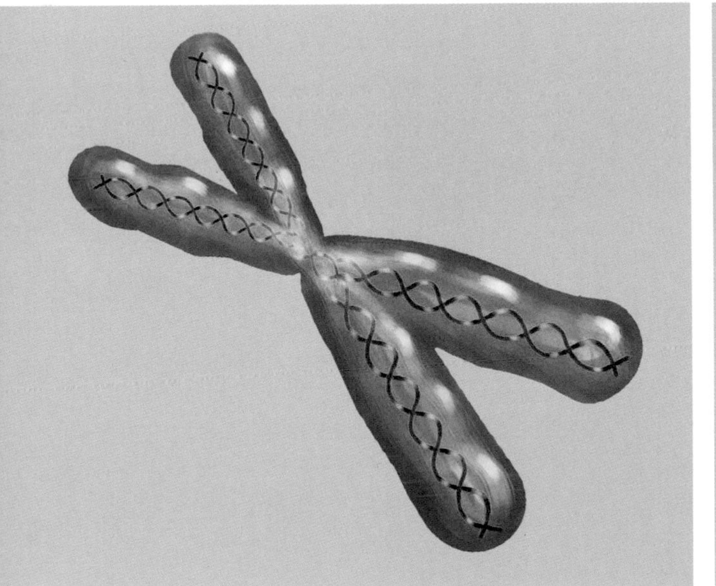

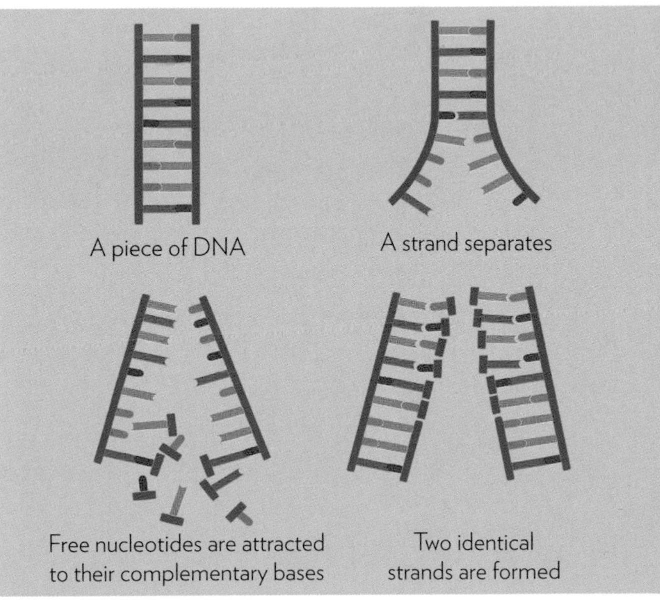

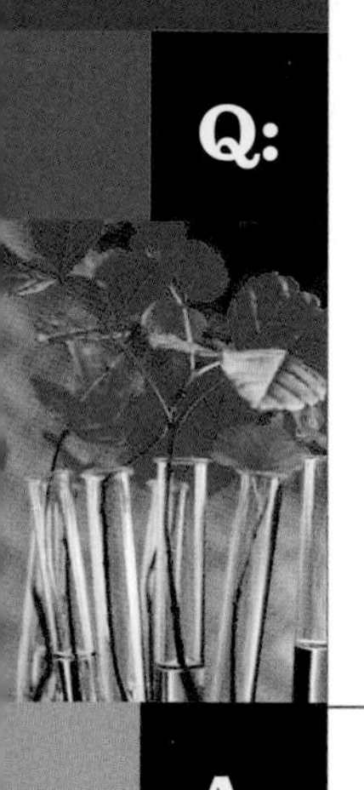

Q:

If we could control DNA, could we bring back any extinct animal we wanted?

Since a complete DNA strand provides a full set of genetic instructions to make an animal (or a plant), it must be theoretically possible to reconstruct extinct animals from their DNA. But could we actually do it?

A:

Current cloning technology can't build an entire animal from a single strand of DNA, but the theory is sound. If we figure out how to reliably insert DNA into "blank" cells, we can make extinct animals. But it turns out DNA doesn't last all that long ....

Think the instructions for flat-packed furniture are complicated? They're nothing compared to DNA. With more than 400 million so-called nucleotides providing instructions for how to assemble a species, a DNA molecule is seriously big.

And any gap in the strand, even a tiny hole, will make the strand useless. Extracting a complete, undamaged strand of DNA from a long-dead animal is incredibly difficult.

Movies and science fiction have suggested we could use the DNA of a living animal to plug up the gaps. The obvious example is to use African elephant DNA to complete a strand of wooly mammoth DNA.

The problem with this, even if it did work, might be philosophical rather than practical: is the resulting animal a real mammoth—or is it just a mutated African elephant with hair?

There's an even bigger problem with using DNA to resurrect ancient species. Chemically, DNA is just a hydrocarbon, a gigantic molecule made of hydrogen, nitrogen, carbon, oxygen, and phosphorus. Compared to something like rock or metal, it's very unstable and delicate.

DNA breaks down by itself over time. Our current models suggest that it completely degrades over about seven million years, though a strand would become useless for cloning before then.

That might seem like a long time, but if seven million is the limit, that means we might never be able to bring back dinosaurs—the last dinosaur died around 65 million years ago.

The most promising extinct candidates for DNA recovery are animals that have gone extinct only recently. High-profile examples include the Tasmanian tiger or thylacine, a large marsupial that went extinct in 1935 at an Australian zoo. There's also the famous Dodo bird from Mauritius, and the Yangtse River dolphin or baiji from China.

These animals are good candidates because museums have preserved specimens, and in the case of the thylacine, that includes fetuses. There's a better chance that scientists could patch together a complete DNA strand from these recent samples.

In fact, in 2008 scientists managed to inject a mouse fetus with a gene from the thylacine responsible for forming bone. It didn't make the mouse look like a Tasmanian tiger, but the research team was able to detect the gene in the resulting mouse fetus. It's a long way from a baby thylacine, but it's a start.

The important thing about DNA is the information it carries—the instructions for building a life form—rather than the actual molecule itself. Assuming we make huge advances in biology and medicine, it's theoretically possible to build synthetic DNA from what's called the "genome"—a detailed description of DNA strand.

Genetic researchers analyze the DNA of living animals to map—or "sequence"—their DNA. This information, which adds up to about 3.2GB for a human, can be used to figure out if a person is carrying the gene for, say, aggressive breast cancer.

But it could also, theoretically, be used to clone the person. Or the sheep. Or the wooly mammoth.

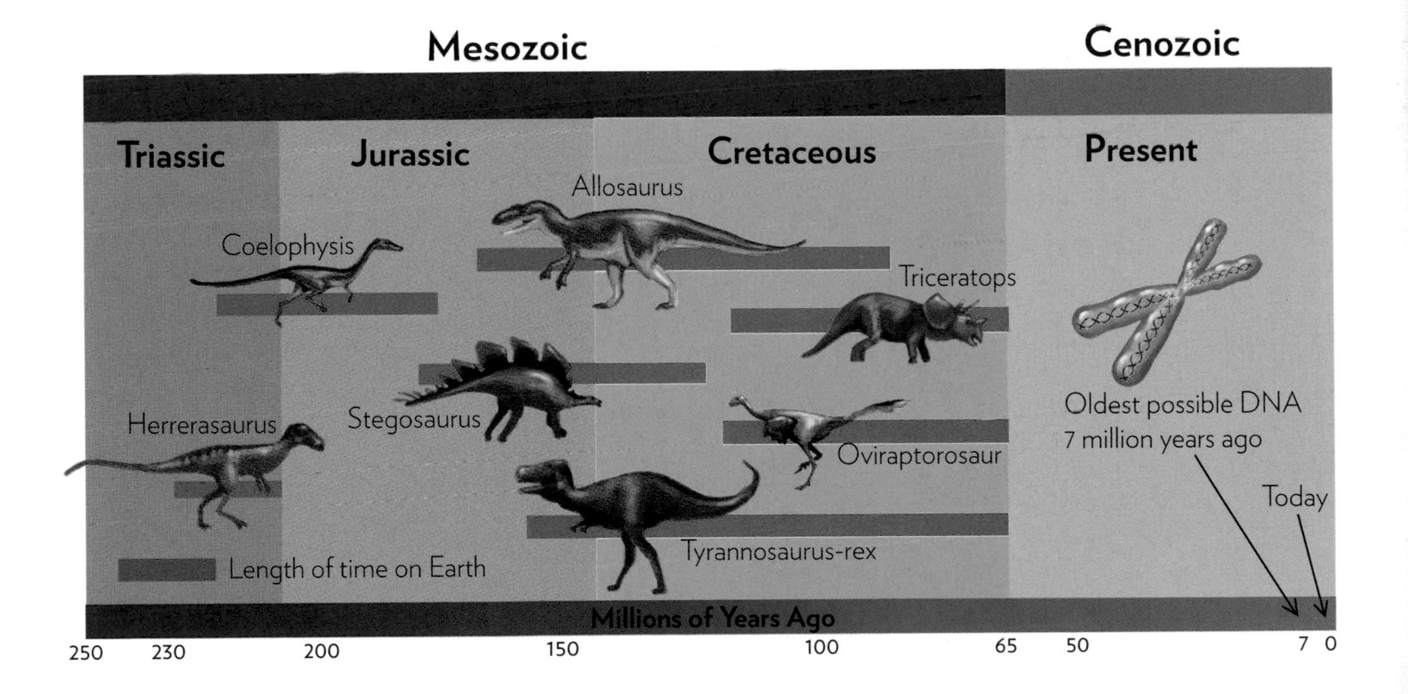

Q:

Why do viruses make us sick ... but only some viruses?

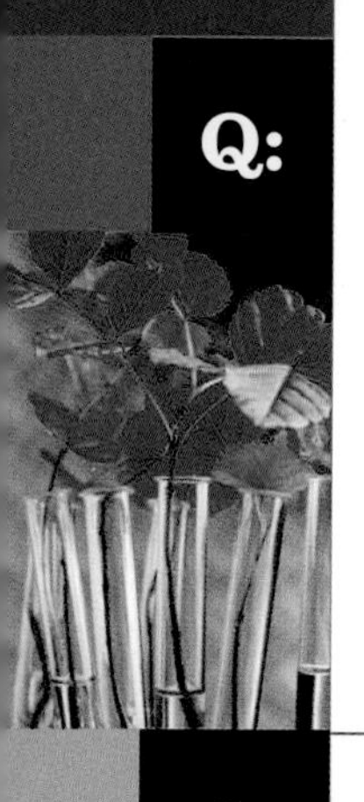

A virus is a microscopic life form that doesn't play by the usual rules. It doesn't have DNA and relies on the cells of a host to reproduce. But since viruses rely on us to live, why do they make us sick ... and even kill?

A:

Some viruses live in peace inside us, reproducing at a rate that doesn't disturb us. But others go crazy, replicating nonstop until they overwhelm our cellular machinery and stop our organs working properly.

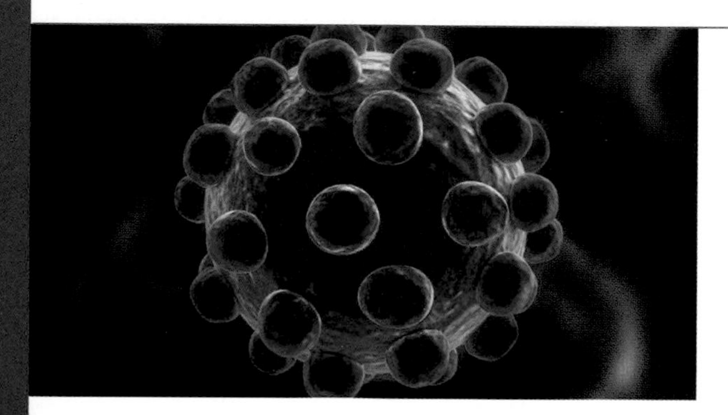

Over time, we have struck a deal with the viruses that live on Earth. If a virus promises not to make us sick or damage our cells, we're happy to let it live inside us. Humans—indeed, all life, even bacteria—carry billions of viruses around in their body every day. We have so many different viruses inside us, scientists are discovering new species all the time.

But some viruses break the deal. They reproduce too rapidly or make so many copies of themselves it disrupts the normal function of our cells. Then, it's war: our immune system does its best to kill off the virus to prevent further damage.

Viruses are strange because they don't have a cell structure like all other life forms. Some don't have DNA, some do. They come in many unusual shapes—some even look like tiny moon landers, legs and all. They're too small to be seen through a normal microscope, but they're the most numerous biological entity ... though since they don't have cells, technically they might not even be alive!

Viruses rely on the cells of other life to reproduce. They enter the cell and replace its genetic material with their own. Instead of making more cells, the cell then starts making viruses.

Many viruses live in balance with their hosts. They only take over a few cells, and only make a few copies of themselves. Sometimes, though, a virus evolves to exploit its host—it rapidly makes trillions of copies, literally exploding the cells of its victim. When enough cells are damaged, the host gets sick and can even die.

We have a defense against these types of viruses—our immune system. Specialized white blood cells hunt down and kill viruses, and the body can secrete a substance called interferon that stops viruses reproducing. A warm-blooded animal like a human can even increase its body temperature so the extra heat shuts down the virus.

Mostly, the symptoms you feel when you have the flu are caused by your immune system trying to kill the flu virus. The fever is the temperature increase that the body hopes will kill the invader. The snotty nose is to trap the virus, and coughing and sneezing expel it from the body. The aches and pains are the buildup of fluids in the joints from the immune system transporting materials.

Unfortunately for some people, the flu can be fatal. The virus especially likes infecting our lungs, and a combination of the damage it does to lung cells and the swelling and fluid buildup from our immune response can kill.

Meanwhile, our digestive systems are full of viruses. Careful analysis of human feces has shown hundreds of different species that never make people sick—simply because they never start reproducing out of control and damaging cells.

Why do viruses exist? Because there are no fossilized viruses, scientists don't know exactly how they evolved. They may be an inevitable by-product of the biological systems that created cells, genes, and the processes necessary for our kind of life. They may even have come from outer space, hitching a ride on comets that hit Earth.

While viruses still kill thousands of people every year, they may also be the key to unlocking new forms of medicine and gene therapy. Their ability to insert new genetic material into a living cell could even lead to a cure for cancer.

Virus attaches itself to a cell

Virus penetrates the cell membrane and injects its DNA or RNA

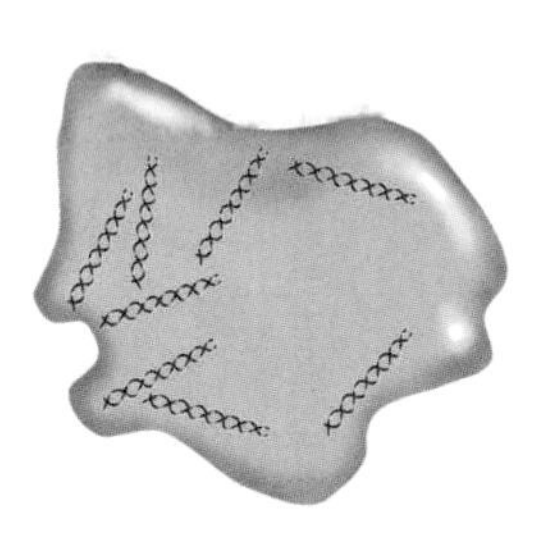

Virus's nucleic acid replicates using host cellular machinery

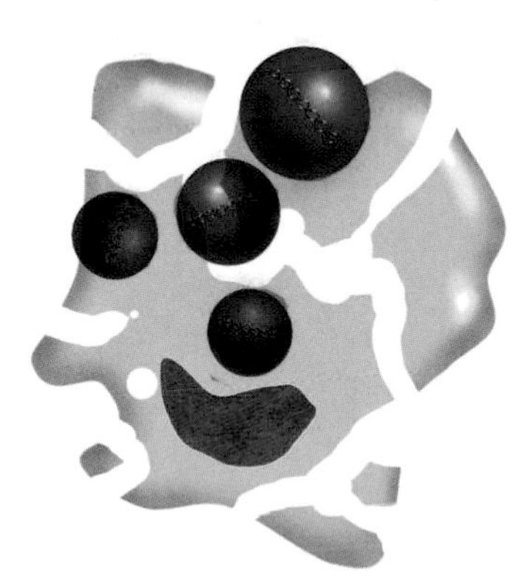

New nucleic acids are put into viral particles and released, sometimes destroying the host cell

Q:

Why do living things age and die?

The longer an organism lives, the greater the chance it will grow weak and eventually die. But if we can recover from diseases and severe injuries ... why do we get old and infirm?

A:

The majority of our cells wear out and need to be regularly replaced. But cells don't divide forever, errors occur during reproduction, and eventually our bodies stop working. But why does this happen?

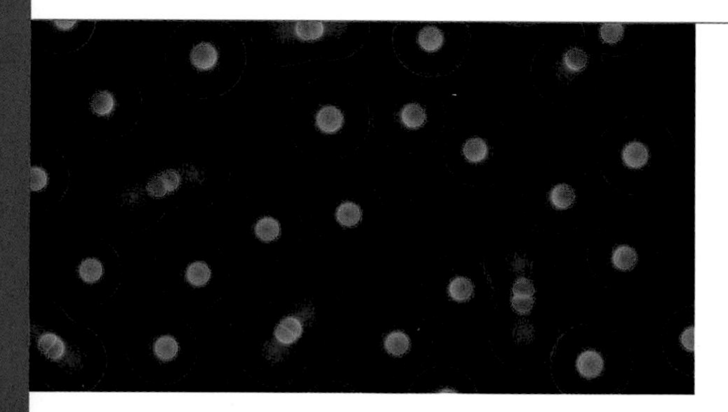

There seems to be an odd contradiction in the way our bodies work. On the one hand, if we cut ourselves, we have the ability to repair the damage with only a scar to show for it. But on the other hand, the longer we live, the more our bodies gradually break down. If you escape disease or accident, you're still doomed to die—most often because the heart or another organ stops functioning.

Today, about two thirds of all deaths are from old age. Scientists call the process of aging *senescence,* and there are two main types.

The first has to do with an organism's cells. As cells begin to wear out, they make copies of themselves, or divide, and the new cells carry on doing the job of the worn-out cells. For reasons not fully understood, this only happens about 50 times for each cell. This might be because the DNA in each cell doesn't copy absolutely perfectly, and over time a part of the DNA called a *telomere* becomes shorter. It's almost like a wick or fuse slowly burning down over the lifetime of the animal or plant.

The second type of aging or senescence applies to the whole organism. It means the body gets worse at doing its job of keeping itself alive. And with cells dying and not being replaced, things only get worse.

Eventually, the system gets so out of whack that even otherwise healthy people develop heart or liver or kidney diseases and ultimately die.

The question of why organisms age is still being worked out. There are many theories: DNA replication isn't a perfect process, and errors creep in. So-called "free radicals"—little pieces of chemicals that react with elements in a cell—cause damage. Oxygen itself causes damage, and in a way, we "rust" to death as our tissues slowly oxidize!

However, evolution has done a pretty good job of eliminating things from our design that disadvantage us as animals. A human who lives longer can potentially have more children and pass on more genes—so why do we still age? Why hasn't evolution favored immortal humans?

Well, that job might be one that's only partly finished. Until very recently (less than 200 years), our average life span was around 45 years. Evolution has had no time to "fix" diseases that you get after age 45—especially cancer.

Some scientists believe that the way aging slows down cell division in your body can be a very good thing: it massively reduces your chance of getting cancer.

This might seem puzzling given how many people get cancer these days, until you look at how old today's "average" cancer patient is—over 50.

It might be that without aging, complex organisms get cancer too quickly, which, ironically, shortens their life span and prevents them from reproducing. The single-celled organisms of three billion years ago that didn't age all died out! For humans, we still get cancer when we get older because evolution has not had time to react to our technology-driven extension of life.

As for us humans, has there ever been another species on Earth that has doubled its life expectancy in a matter of decades? Probably not. Now the race is on to see who can "cure" aging: nature, or us!

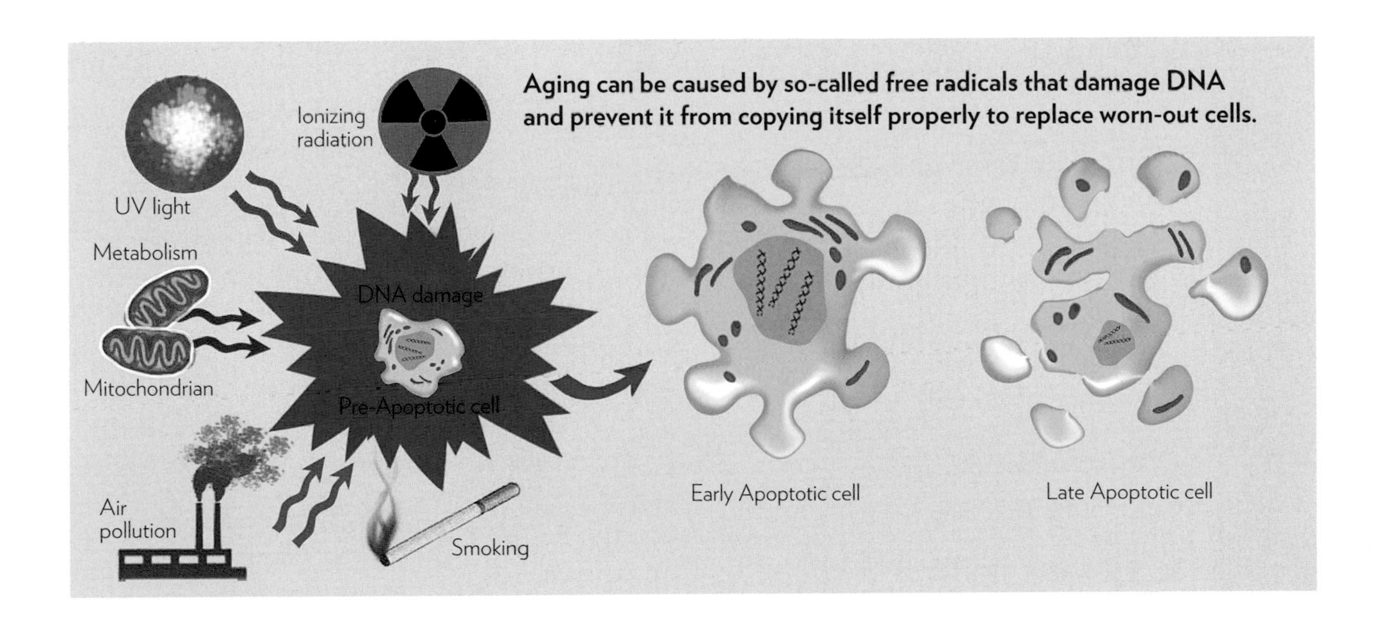

Aging can be caused by so-called free radicals that damage DNA and prevent it from copying itself properly to replace worn-out cells.

Q:

Do all living things die? Are there any immortal species?

Given the huge variety of life on Earth and the way species have adapted to so many niches and taken so many forms, surely there's at least one organism that is immortal.

A:

All cells die eventually, but some species have indeed evolved the ability to live more or less forever ... but you do need to use a very careful definition of "immortal" ....

The answer to whether there are any immortal life forms on Earth is actually quite tricky and really depends on how you define both the idea of "living forever" and also the idea of what makes an individual life form.

There are various species of plant and fungus that have extremely long life spans. Plants are interesting because they age quite differently from animals. Animals age as their individual cells lose the ability to divide and create new versions of themselves (see previous pages for more). The cells wear out and die, and the animal loses so-called biological and metabolic function.

When certain types of plants lose cells due to age, the plant overall gets tougher, and its remaining cells become more efficient. It can pump water higher up into the plant—this is why older trees can be so tall. As long as the tree is not damaged by disease, insects, or storms, there's no biological reason for it to die—at least within the timespan we've known about this ability. This only applies to certain groups within the plant kingdom, though—other plants are genetically programmed to die every year, leaving seeds behind for next spring.

Trees like the aspen lead two lives: one above ground, and one below. The trunks and leaves live only 40 to 140 years, but their root systems are ancient. Some are estimated to be over 80,000 years old.

Then there are weird creatures like the tardigrades, microscopic critters called water bears that can go into suspended animation and survive, neither alive nor dead, for years.

There are even certain jellyfish that can reverse their aging, and actually dismantle their bodies back into their immature form so they can start growing all over again.

And vast fungus colonies that live underground have been tested at many tens of thousands of years old.

The problem with all these examples, though, is that the older an organism is, the more likely it is to have "cheated" at immortality. It's not like the actual plant itself has stayed alive, but more that it has cloned itself. In the case of long-lived trees, a shoot will grow into a new tree, remain connected to the old tree, and the old tree dies. Is the new tree the same tree? It is genetically identical, but is it the same individual? Scientists tend to stick with strict biological definitions of immortality, and in this case, that plant is considered immortal.

The other problem is that we haven't had the technology to test the age of really old organisms for very long. We may simply not have known about, say, the Rougheye rockfish long enough to know if it will live forever or "only" 250 years.

We certainly haven't discovered any individual organisms that are millions of years old. Immortality doesn't just mean "lives for a couple thousand years"; it means "lives forever."

Humans may be the only species obsessed with immortality. Lots of research is going into "curing" us of aging, and when you bring in ideas like uploading your mind to a computer or cloning yourself a whole new body, then perhaps the first immortal creature on Earth is already here ... us!

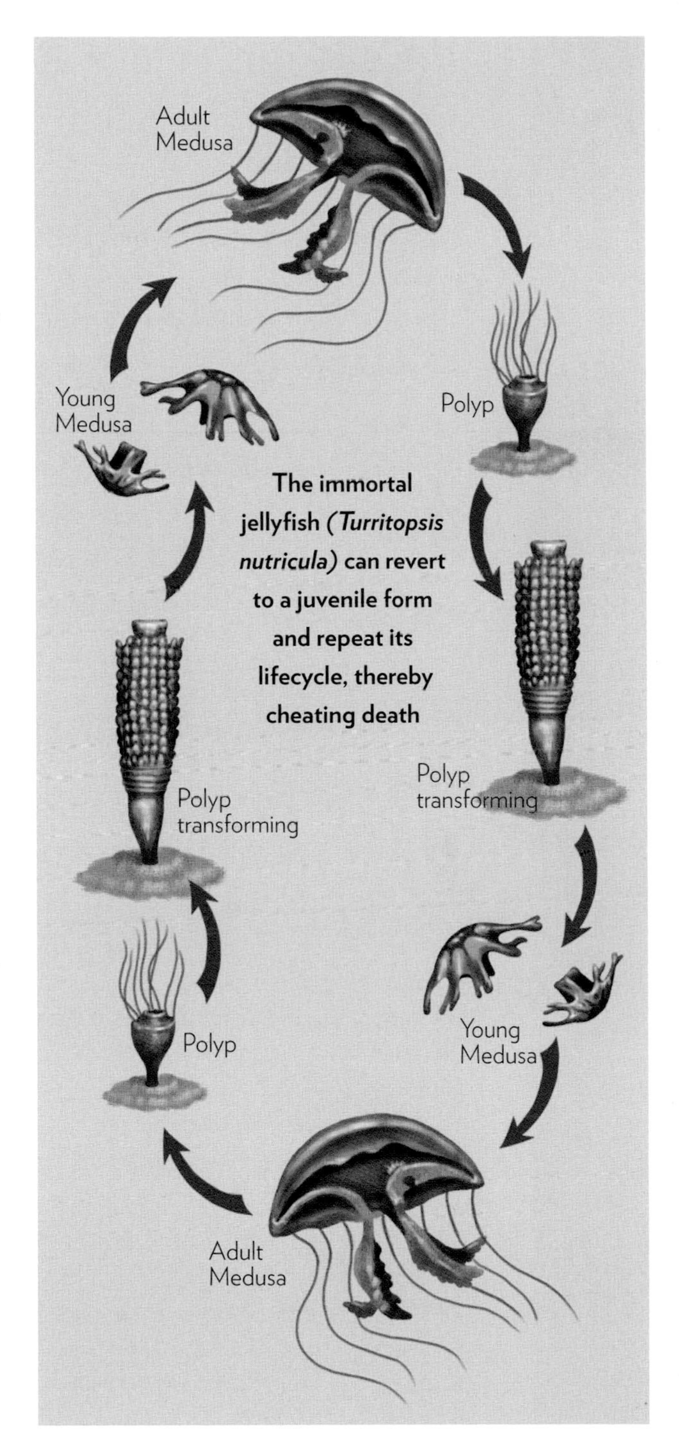

The immortal jellyfish *(Turritopsis nutricula)* can revert to a juvenile form and repeat its lifecycle, thereby cheating death

Q:

Is there any evidence humans are still evolving?

Humans have only been a high-tech species for a very small part of our total history. So surely evolution still applies, changing our bodies and behavior. Is there any evidence of this actually happening?

A:

Humans are absolutely still evolving, changing physical traits rapidly enough for us to see. Our wisdom teeth, our ability to drink milk, and the strength of our immune system are all being affected by continuing evolution. But what happens when our technology takes over?

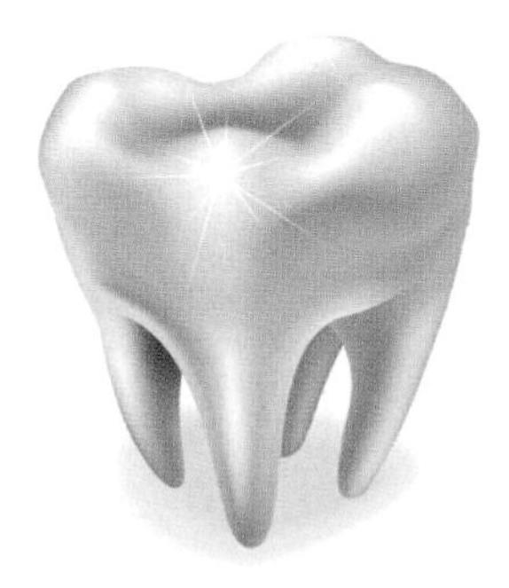

To say evolution is caused by natural selection is only really a summary of the many processes involved. A more complete way to describe evolution is to say it's the way life changes over time to find the best balance between becoming as tough as possible for whatever environment it finds itself in, finding an environment that no other species is using, and also using as little energy as possible to achieve this.

In other words, evolution doesn't just select for the "best" traits, it also selects for the most efficient. The human immune system is a great example.

Today, for an increasing number of people, it's not really necessary to have a super-strong immune system. We've eliminated many dangerous diseases (such as smallpox) and made it so the chances of catching many others are incredibly low.

Why do we need an immune system capable of responding quickly to a polio infection when we give everyone a polio vaccine in childhood? Evolution finds a balance between not having to use lots of energy to maintain a strong immune system that never gets used and an immune system that's still strong enough to respond to the vaccines and other diseases we still get.

Another area of recent evolution is in the loss of our wisdom teeth. A hundred thousand years or so ago, humans had big jaws and needed a set of powerful molar teeth at the back to grind up plants for digestion.

Today we use tools to grind our plants (such as making wheat into flour), so there's no longer an evolutionary advantage to having wisdom teeth. Around 35 percent of modern humans don't develop wisdom teeth. People without wisdom teeth don't have to go through all the fuss and expense of having them removed, or suffer the problems of wisdom teeth crowding into a jaw that's now too short for them. This is a very slight evolutionary advantage, and over a long enough period of time—maybe another 100,000 years—it's likely the majority of the population will no longer grow wisdom teeth.

Another recent adaptation for some adult humans is milk tolerance. Babies can digest lactose thanks to an enzyme called lactase, but our bodies used to stop making the enzyme when we got older. But as little as 10,000 years ago, Europeans started producing lactase in their gut all the way to adulthood. Most humans actually can't drink milk—it's more normal to be lactose intolerant. Over time those people who happened to have the enzyme were more successful and had more children, and those children bred, and so the trait for lactose tolerance was passed down and is now in 35 percent of the population.

These are just some of the ways humans are evolving based on evolutionary processes that have existed for a billion years. But things are about to change: with both knowledge and technology, humans can now direct our own evolution. Some scientists predict that we will make our eyes much bigger so we can live on planets farther from the Sun. Others say we might engineer computers into our actual bodies—specialized cells that let us connect directly to the internet without the need for a separate device.

We might choose to change our bodies deliberately in the millennia ahead. We could make our brains (in blue) bigger, and as we colonize space we might give ourselves larger eyes (because the light is dimmer farther from the Sun).

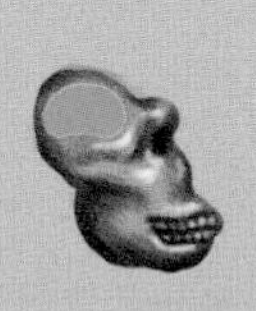

Singe anthropoïde

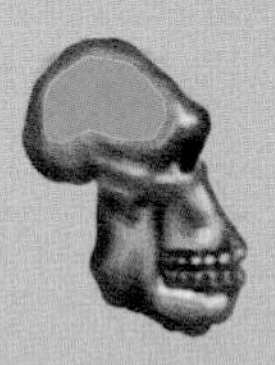

Australopithecus africanus

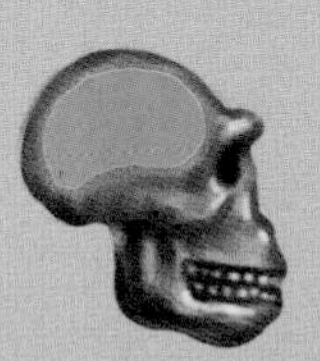

Homo habilis

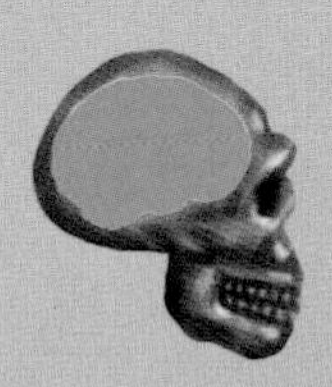

Homo erectus

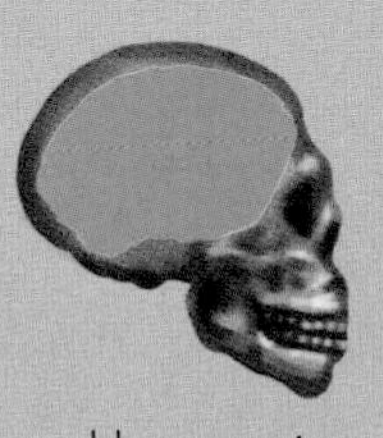

Homo sapiens neanderthalensis

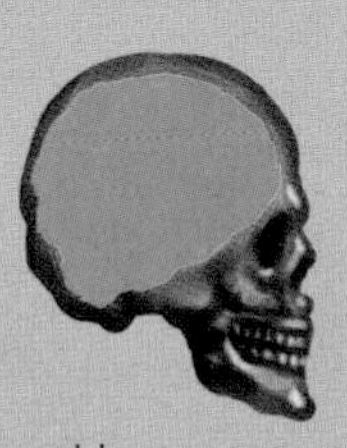

Homo sapiens sapiens (modern man)

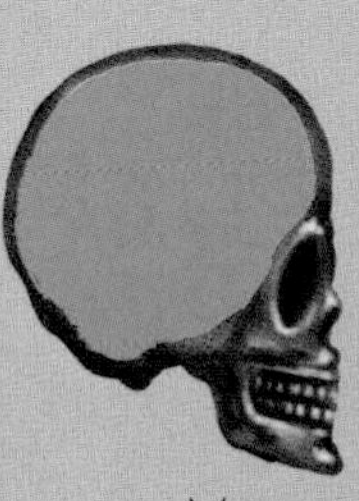

Man in the future

Q:

Why are there so few kinds of large mammals?

Flip open any dinosaur book, and you'll see dozens of different species of really huge animals. But today the number of animals bigger than a human seems really low. What's the explanation?

A:

We are living on a damaged Earth, recovering from an Ice Age and a changed ecosystem. Tougher conditions made it harder for big mammals to survive. But there was one extra major factor in their extinction ....

Pick a random living species from the whole mix of nonmicroscopic animals, and odds are you'll get a beetle. When it comes to species diversity, smaller critters have us big critters beat.

Sure, the largest animal to ever have lived—the blue whale—is alive right now, but this question isn't about breaking single records, it's about why there are comparatively fewer types of big animal than in the past.

Looking at the fossil record, there seems to have been hundreds of different species of big dinosaur. There's also evidence of many species of very large land mammal, too, capped off by the mighty Paraceratherium. This giant hornless rhinoceros would have dwarfed a modern elephant—the biggest individual we've found is estimated at 30 tons!

If you visit a dinosaur museum these days, you'll probably find a new gallery dedicated to extinct mammals from the last few million years. After all, the dinosaurs died out around 65 million years ago, so there was a *lot* of intervening time between them and us for evolution to experiment with other life forms.

And indeed the fossil record shows many species of large mammal, from pig-sized elephants to strange tusked things that don't really look like anything alive today. There were wooly versions of many modern mammals—mammoths, of course, but also wooly rhinos.

So what happened? Well, first let's get rid of a misconception. Because our first encounter with dinosaurs tends to be in children's books, we see twenty or thirty species all at once. But in fact the amount of time that passed between the extinction of the Stegosaurus and the evolution of the Tyrannosaurus rex was greater (83 million years) than the time between the T. rex and us (only 67 million years). In other words, if you travelled in time to the dinosaur era, there would be far fewer types of large dinosaur alive at any one time than you might expect.

The same applies to mammals—across the whole fossil record, the number of large mammals is pretty respectable.

But there are definitely reasons there are fewer large mammals alive today. The first reason was a series of Ice Ages. Cold conditions slowly eroded the biodiversity of mammals. But the really big impact came after the last Ice Age, about 12,000 years ago.

Already weakened by massive changes in sea level and global temperatures, many species of large mammal went extinct. But they'd survived previous Ice Ages—what was so different about this one?

The difference was the arrival of a new kind of predator. Individually, it wasn't much of a threat, but it could band together into packs of formidable hunters. What's more, evidence suggests this predator targeted young members of a herd, which for slow-reproducing creatures meant no new mothers to bear the next generation.

The name of that predator? You've probably guessed—it's us. Humans became some of the most effective land hunters ever, and we quickly adapted to being able to hunt in the sea. Whales, elephants, seals, rhinos—nothing big was safe.

Whether humans are solely responsible for the loss of biodiversity in big mammals is still the subject of debate. But signs point to us being the culprits. What the ice started, humans finished.

Q:

Why aren't there any half-evolved animals?

One of the puzzles of evolution is that even though animals are supposedly evolving all the time, every animal alive seems perfectly adapted for its habitat. Where are all the animals that are only halfway finished evolving?

A:

Evolution doesn't come with an easy scoring system. Animals aren't more evolved or less evolved—they slowly change as their environment changes. Even so, there are a few quirky beasties that certainly ***look*** half evolved ....

Evolution can be tricky to get your head around, because the timescale across which it occurs is so huge. What does a million years mean, really? It's an unimaginable span of time. The whole of human history—including recorded history, history with archaeological evidence, and theoretical prehistory—is too short a span of time for us to show evolutionary changes except in a few very minor areas (see "Is there any evidence humans are still evolving?").

At its most basic level, evolution occurs when random mutations happen and give the life form some kind of tiny advantage over others of its species. We're not talking two individuals battling to the death, but rather subtle trends. Over millions of years, they add up—eventually so many that the animal is unrecognizable. It becomes a new species.

The thing is, individuals within a species of animal (or plant) aren't identical to each other. This is obvious in humans: some of us are taller, some have have bigger teeth, some more widely spaced eyes. If the environment changes so that people with widely spaced eyes produce, say, 0.01 extra children each, then after many tens of thousands of years there will be more wide-eyed humans than narrow-eyed. If the trend continues, then "wide-eyed" becomes the "normal" trait.

Strictly speaking, all life on Earth is the same. There are some fundamental divisions between life forms with a nucleus in their cells (eukaryotes, which we are) and simpler life without a nucleus (prokaryotes), but we all evolved from a common ancestor of some kind.

The idea that, for instance, the Australopithecus (a human ancestor) wasn't "finished" evolving until it became a Homo sapiens (us) is kind of wrong. The Australopithecus was well adapted to its environment. It was the species of the day.

Each new generation is subtly different from the one that came before it. Eventually enough generations pass that if you compare an animal from Generation 1 to an animal from Generation 1,000,000 you can see they are very different—so different they can't even breed. But there's an unbroken genetic chain between them of a million mothers giving birth to a million babies.

Still, it can be fun trying to spot animals that look like they're only "half evolved." There are some good examples.

Some people who argue against the theory of evolution point to whales and say that if whales really did evolve from land-dwelling mammals, there would be evidence of some kind of halfway-whale. It would still have fur and whiskers and probably a rather doglike face. It would have flippers, but they would still have fingernails, and the rear fluke would be a pair of feet fused at the heel. And these creatures would have to breed on land and only hunt in the sea.

This creature exists today—it's called a seal. And it's good evidence of a land-based animal evolving into a sea-based one. After another million years, the descendants of seals might look more like dolphins—smooth skin, no whiskers, rigid flippers, and no need to ever return to land.

Then there's the flounder and sole—fish that are born normal, but as they age one eye migrates across the top of the head so both eyes are on the same side and the fish can lie flat on its side on the seafloor. They all look like mutants!

A Partial List of the Predecessors of Today's Whale

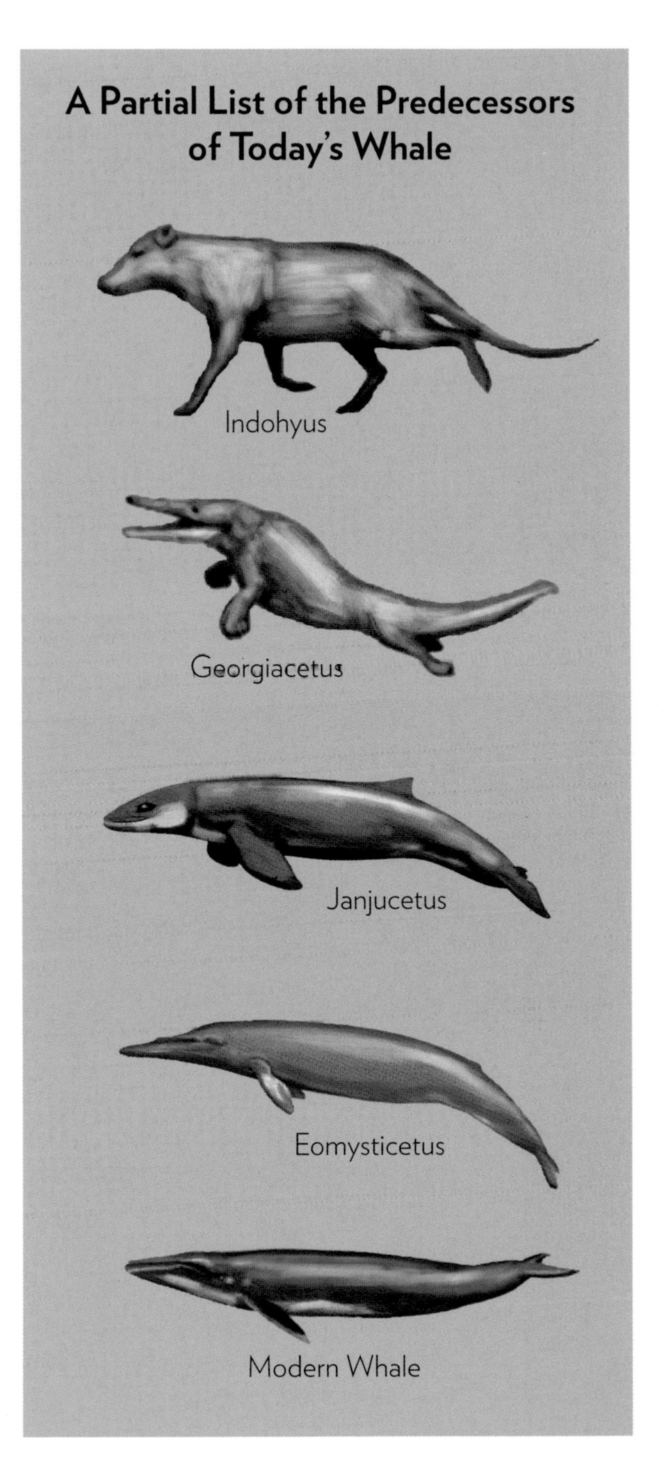

Q:

Why are some animals poisoned by foods that are harmless to humans?

Our pets can be killed by the same sorts of poisons as us—especially snake and spider bites—but there are also many supposedly harmless substances that can kill our furry friends. Why does this happen?

A:

The shocking truth is that many of the things we eat that we think aren't poisonous ... actually are! They're full of natural pesticides and other toxins. It's just that humans are large animals, and picky eaters. Our pets, on the other hand ....

All responsible dog owners know not to let their pooch gobble down a big block of dark chocolate. And cat owners know aspirin tablets can be fatal to their kitties. The list of substances harmless to humans but dangerous to animals is long. For dogs alone it includes avocados, macadamia nuts, grapes, and various artificial sweeteners.

A poison is a wide-ranging term for any substance that can have a negative health effect (usually on a human) because it causes some change in the chemistry of our metabolism. It might crash our blood sugar levels, shut down our livers, paralyze our lungs, or cause a heart attack.

Scientists also use the terms "toxin" and "venom" when talking about specific types of poisons. A toxin is a substance deliberately made by a plant or an animal inside its tissues that will kill a predator if the predator eats it. Toxic plants and animals often advertise—usually with bright colors—to warn predators against eating them.

Venom, on the other hand, is a toxin that can be injected into another animal. Snakes and spiders are the most infamously venomous creatures—their sharp fangs work as syringes, injecting the venom into prey—usually for hunting (snakes), and sometimes for defense (some frogs).

The funny thing is, chocolate and avocados are actually toxic, full of stuff designed to stop the plants they come from getting killed. Avocados have a fungus-killing substance in them called persin. And chocolate—made from cocoa beans—contains caffeine, which is an insecticide.

These two toxins have a limited effect on humans. Caffeine, as we all know, does affect us, but in a way we see as positive. An increased heart rate picks you up in the morning, but for a dog—a smaller animal with a resting heart rate as much as 100 beats per minute—a big dose can lead to a fatal heart attack. The risk is compounded by the fact that chocolate has additional caffeine-like substances in it, particularly methylxanthine and theobromine. Also, the amount of sugar and fat in a large dose of chocolate (dogs do tend to gobble the whole lot at once!) can even crash their pancreas. That's not necessarily fatal, but can be very painful.

You may have heard the old wives' tale that putting out a bowl of milk full of crushed-up aspirin will take care of a pesky stray cat. Well, it's no myth—aspirin in large doses will definitely hurt or even kill a cat via hepatitis, gut inflammation, and even respiratory failure. But paracetamol is even more dangerous. Cats can't flush it out of their systems; it just hangs around and damages their liver, gives them jaundice, and even destroys their blood!

At the end of the day, human curiosity and intelligence lead us to eat a whole lot of things that nature spent millions of years filling with poison. We know to eat only small amounts, or how to prepare them to break down the poison. Animals, on the other hand, just gobble everything they can get their paws on. A couple grapes on a hot summer afternoon is delicious for us. A whole bagful inhaled in 18 seconds by a dog can be fatal.

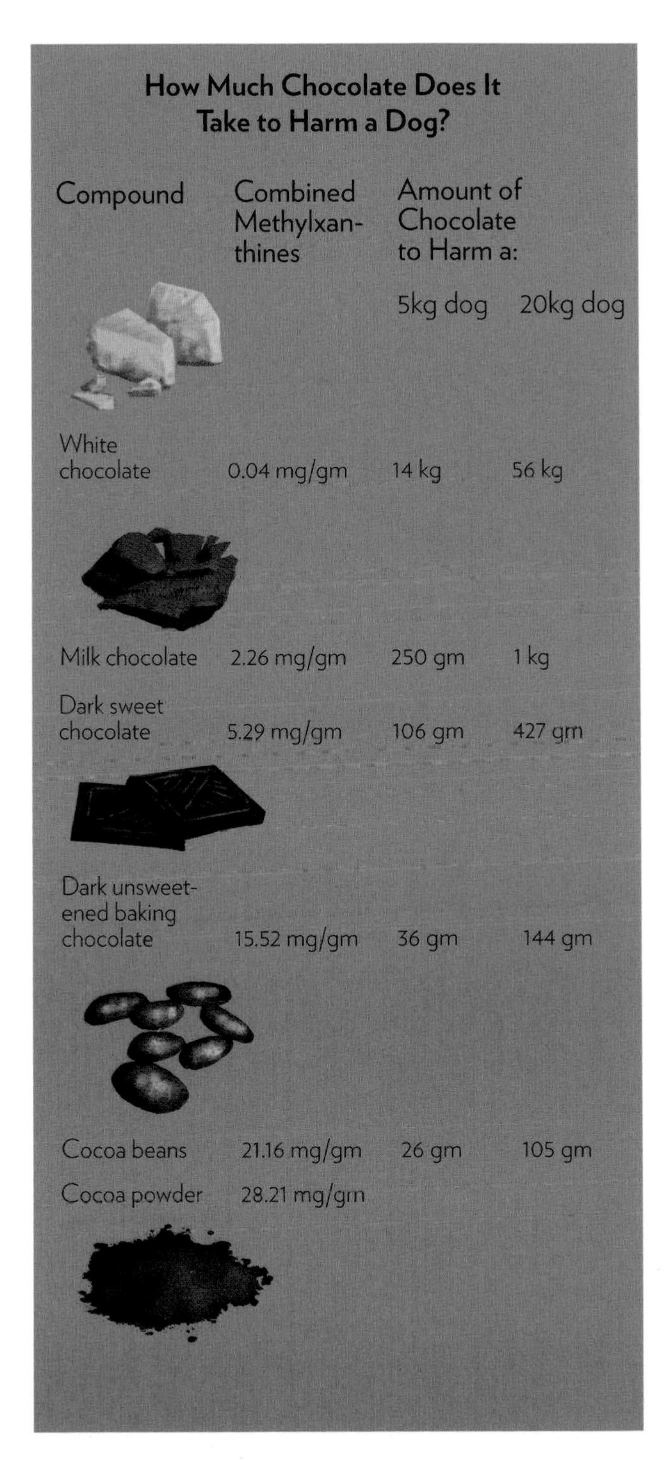

How Much Chocolate Does It Take to Harm a Dog?

Compound	Combined Methylxanthines	Amount of Chocolate to Harm a: 5kg dog	20kg dog
White chocolate	0.04 mg/gm	14 kg	56 kg
Milk chocolate	2.26 mg/gm	250 gm	1 kg
Dark sweet chocolate	5.29 mg/gm	106 gm	427 gm
Dark unsweetened baking chocolate	15.52 mg/gm	36 gm	144 gm
Cocoa beans	21.16 mg/gm	26 gm	105 gm
Cocoa powder	28.21 mg/gm		

Q:

Why can't birds taste chili peppers?

Unlike mammals, birds can gobble as many crazy hot chilies as they like, straight from the bush. They simply don't feel the burn. Why not, and why is the chili plant so selective about whom it tortures?

A:

Birds lack the taste receptors on their tongues to feel the burning sensation from chili. Humans do have the receptors and can feel the pain. Lots of pain! But ironically, chili may have a medical role to play in pain relief ....

One of the fascinating things about the plants we eat is that many of them contain toxins the plant has spent millions of years evolving—just so animals like us won't eat them!

Only humans seem to be perverse enough to actively seek out plants that actually hurt us to eat. Little does the poor chili realize, but the correct amount of capsaicin (the chemical that makes the burn) actually enhances the flavor of carefully prepared meals. But only because humans are crazy!

As anyone who has been naughty or unlucky enough to be hit by capsicum spray will agree, capsaicin burns any tissue it comes into contact with. The plant is definitely sending us a message: don't eat my fruit!

Birds don't get that message. Their taste buds don't react to capsaicin. It simply doesn't register in their mouths, and so birds can happily eat chilies with no ill effects. Later, the bird flies off, poops out the seeds, and a new chili plant germinates. Everybody wins!

We mammals, unlike birds, have a nasty habit of chewing our food, and our powerful back teeth grind up and destroy many seeds. Chili plants evolved capsaicin in their seeds to discourage mammals from eating the fruit. Plants without capsaicin got munched by ancient herbivores and didn't propagate as widely or successfully as those with spicier seeds.

Humans have learned to love the burn of a good chili. As far as we know, we've always enjoyed spicy food. Part of the explanation may have to do with the way the brain releases endorphins as the burn of the capsaicin fades.

Humans actually compete with each other to produce the most powerful chili-based concoctions, ranking them on the so-called Scoville scale. Tabasco® sauce has a rating of 2500–5000, cayenne pepper 30,000–50,000, habanero chili 100,000–350,000, and the Trinidad moruga scorpion (a new variety of chili, not a killer arachnid) tops out at a face-melting *two million.*

One step up from crazy foods, we use capsaicin as a nonlethal weapon, spraying it in the eyes of rioters or rowdy criminals in hope the tears and pain will convince them to mend their ways.

Oddly, though, hyper-concentrated capsaicin can also be used as a painkiller. It works basically by overloading pain receptors so you don't feel pain while the capsaicin is on your skin. Note: you'll get a topical anesthetic from a nurse first. Without it, this particular cure would be worse than the disease.

It's also possible that capsaicin has an important role to play in reducing cancer tumors, and even curing leukemia.

Incidentally, chilies feel hot because the capsaicin causes certain pathways in your pain receptors to open. These pathways normally don't open unless your skin gets very hot—114°F, to be precise. Capsaicin makes the channel open even when your skin is normal body temperature, which is why you get the false sensation of real heat on your skin or tongue if you come in contact with chili peppers.

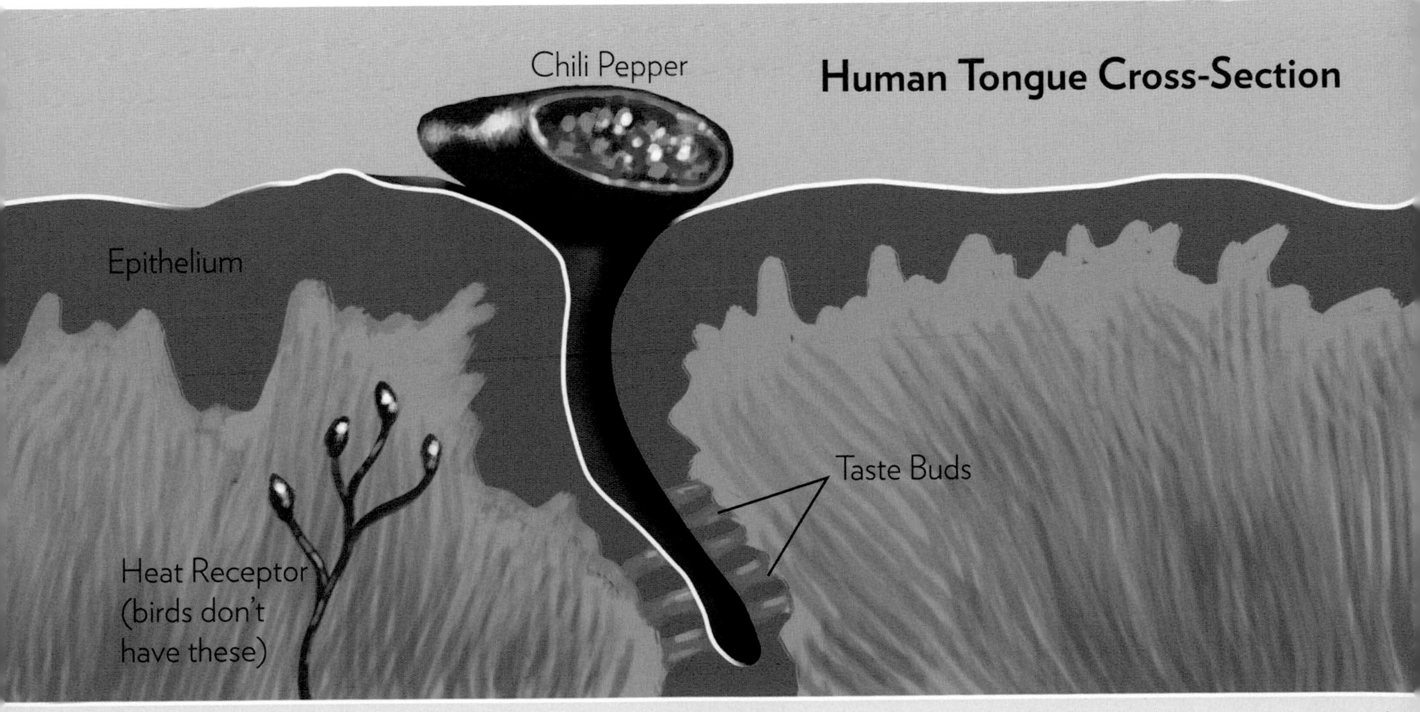

Q:

What makes spider silk so amazingly strong and light?

Spider silk has such amazing properties of strength, lightness, and flexibility it makes human engineers jealous. So why is this stuff so incredible?

A:

Spider silk is made of protein and has what engineers call "exceptional mechanical properties." It's not just strong, it's also very stretchy. The secret? Special glands in the spider that "assemble" the silk.

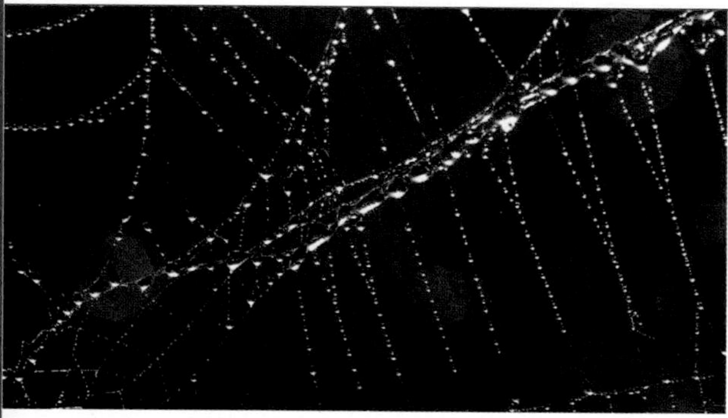

As humans gradually learned how to smelt metals and dress stone, building stronger and stronger structures to protect us from the elements, little did we know that the humble spider was spinning a material that, to this day, outperforms almost all of our most sophisticated creations.

If you walk through a really big web, you might get a sense of the strength of spider silk. Even though this structure made by a tiny arachnid is barely visible, you actually need to exert quite a bit of force to push through it. What's more, a web normally breaks where it's anchored to plants or objects—it wraps around your face and you have to pull it off. Usually while shrieking.

Spiders make different types of web to do different jobs, from the famous sticky fibers to catch insects (capture-spiral silk), to amazingly strong "guy ropes" to hold the web up (major-ampullate or dragline silk), and a super-tough version for wrapping up prey (aciniform silk). They can even make incredibly thin strands of gossamer that baby spiders use to fly to new hunting grounds in a process called "ballooning."

We talk about spiders "spinning" silk because it does really look like they are spinning the silk from their bodies—sometimes they even gather the silk with their back legs similar to a human working a spindle. But in fact, spider silk is made in a process called "pultrusion," where the force of pulling the silk material out of a gland full of pre-silk goop forms it into a thin strand. Spider silk is unique because almost all other biological fibers are made by smooshing material together, whether it be keratin (like in our hair) or even poop. Spiders can also eat and reuse their silk.

Think that's cool? Once mechanical engineers started analyzing spider silk in detail, things hit a whole new level ....

These tiny, often transparent strands have our toughest materials beaten hands-down. By weight, spider silk is five times stronger than steel, and ten times tougher than Kevlar—which is used to make bulletproof vests! It can stretch to five times its length before breaking. It can hold that strength between -40°F and 428°F; and if you put it in water, it contracts by 50 percent.

All these properties make it ideal for human uses. But we don't fully understand how it's made, and attempts to produce artificial silk, while improving, still have a long way to go.

So why not just farm spiders? We can certainly "milk" individual spiders for silk. But there's a problem: unlike silkworms, if you put a whole bunch of spiders together they usually just kill each other. They are, after all, territorial predators.

We're not giving up, though. Spider silk, or an artificial fiber derived from it, would change the face of human engineering. It's a prize worth working for.

Q:

Why can't animals make energy from sunlight like plants?

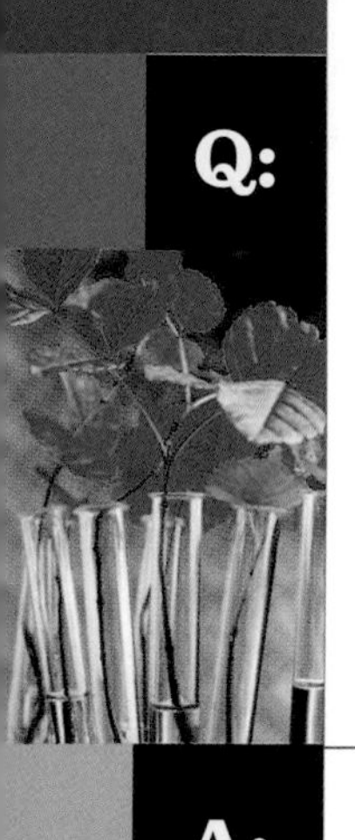

Being able to get a little extra energy from sunlight sounds like it would be a good idea for animals, especially through tough times. So why don't any animals do it?

A:

Sunlight actually provides very little energy, and carrying around the ability to photosynthesize just isn't worth it for animals. Though that hasn't stopped some species trying ....

Ever been hungry and looked at a plant and thought, that guy just gets all his food for free from the Sun—I wish I could lay back, soak up some rays, and feel refreshed and re-energized?

Photosynthesis, the ability to extract energy from sunlight, is an amazing adaptation that solves a big survival challenge for plants: how to get enough food when you're stuck in one place for your whole life.

But it turns out photosynthesis isn't that great. You need to grow lots of leaves so you can have a massive surface area to catch the most rays. And even then the Sun doesn't provide you with much energy at all—at least, not compared to the sheer bulk of calories consumed by an animal every day.

Plants don't move around because they just don't get the energy for it from the Sun. In terms of calories, a plant gets by with far less energy than you do—even a plant that weighs the same as you.

Evolution isn't just about "survival of the fittest." It's also about finding the most energy-efficient way to keep an organism alive. Adding photosynthesis to an animal's ability to extract energy from food just wasn't efficient. The amounts of energy are so small, you would have to stand in the Sun for weeks just to get as many calories as eating a big steak.

By weight, plants use a lot more water than animals—they can be as much as 95 percent water (humans are about 60 percent water). And plants have the luxury of being able to absorb water slowly and constantly through roots and dew. We have to drink.

Animals don't photosynthesize because there's never been an evolutionary reason for them to "eat" sunlight. Though of course, since this is nature we're talking about, there are some exceptions ... kind of.

There are some groups of invertebrates that use sunlight to make food. Well—it's a bit trickier than that. What they do is encourage algae (tiny green plants) to grow inside their tissues. The algae gets a safe place to live, and the animal gets to steal some of the energy the algae makes from sunlight.

The most famous animals to use this system are the corals. Contrary to common belief, the algae in coral is brown. The amazing colors come from proteins made by the coral itself. If water conditions are poor, the coral may stimulate more algae to grow, causing "browning." It's the opposite of coral "bleaching," where the animals expel the algae from their tissues, again in a response to poor water quality.

Giant clams also grow algae in their flesh to get a little extra boost of energy. But both giant clams and the corals have something else in common—they don't move around. Anchored to the seafloor, the extra energy provided by the Sun is worth the trouble of managing all that algae.

Within a hundred years, or maybe even less, it's likely that human technology will emerge as the best photosynthesizer on Earth. Our solar panels can extract huge amounts of solar energy, putting plants to shame. And we turn it directly into electricity—no messing about with sugars!

Q:

Doesn't higher CO_2 in the atmosphere make plants healthier?

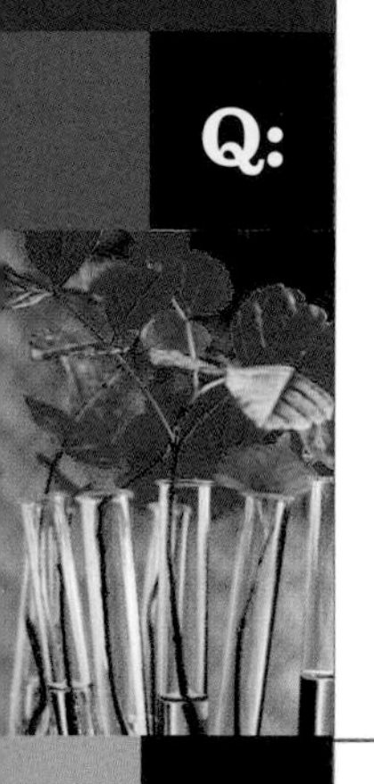

Every grade-school student knows plants take in carbon dioxide and release oxygen. So it seems common sense to think that if there's more CO_2 in the atmosphere, it will help our crops grow. But is that really true?

A:

Yes, plants benefit from more CO_2. But the equation is more complex than that, because the way in which plants respond to more CO_2 isn't always good for us ....

There are a bunch of standard arguments used by people who want to believe that pumping lots of carbon dioxide into the atmosphere isn't necessarily a bad thing. One of them is that plants need CO_2 and will grow more vigorously and healthily in a CO_2-enriched environment.

The simple answer to this is yes, plants do benefit from higher CO_2. And in prehistoric times, CO_2 levels were much higher than they are today. But the issue of climate change isn't primarily about how other life forms will be affected—it's about how humans will be affected.

Plants will benefit from higher CO_2. But will we benefit from the changes that occur in those plants? Not necessarily ....

Photosynthetic organisms use CO_2, water, and energy from the Sun to drive a chemical reaction that makes sugar. Plants then use this sugar for energy. Change the amount of CO_2, water, or sunlight, and the amount of energy changes.

When plants have lots of energy, they grow vigorously. But we don't necessarily want plants—especially crops—to grow willy-nilly. Humans mostly eat the reproductive organs of plants: the seeds and fruits. When we do eat leaves, we prefer young, juicy leaves. We can't digest wood, and we don't much like big, thick stalks with lots of fibers in them.

Unfortunately, extra CO_2 gives plants the energy they need to grow exactly the parts we don't want. Experiments with high CO_2 see plants grow bushier, putting out more leaves and stems, but they don't necessarily make more seeds.

Having to deal with more unwanted plant material will affect the efficiency—and cost—of our agriculture. Farmers will need to process and discard more "waste" matter to get more or less the same amount of grain.

There's another problem: increasing CO_2 only gives the plant the potential to make more energy. To actually make it, the plant will need to match the increase in CO_2 with an increase in water. A more vigorously growing crop will demand more water—and our water supply is already stretched in many places. If farmers don't increase water, the plants won't develop properly and might even end up making *less* seed.

Beyond these basic problems, things start to get more complex. Experiments show that for some reason, insects really like eating plants that have grown in higher-CO_2 environments. Soy plants, in particular, suffer more nibbling from bugs when they've been grown with extra CO_2. It's hard to predict if this will be true for all our crops or just some.

We've already mentioned how many of our food plants produce toxins. The plant needs energy to make those toxins, and with extra CO_2 providing more energy, it's possible the plant will become more toxic. CO_2 could turn your guacamole deadly.

The plant kingdom is a complex network of life, with thousands of different species. All will react differently to increased CO_2. Some will benefit, others could die. Our coffee could get extra caffeine, which might be good. But our wheat might demand more water to grow, which would be bad.

The problem with climate change isn't that the biosphere will collapse. It won't. But it will *change*, and even change that actually benefits some life could have a massive negative effect on us.

Q:

Cheetahs are the fastest, elephants are the biggest ... what's a human's "animal superpower"?

When it comes to physical excellence, humans don't seem to stack up that well. Oh sure, we're really smart and can build machines to beat animals at everything—but that's cheating. Is there anything we're "world's best" at naturally?

A:

Humans have a number of unique physical adaptations that make us extremely adaptable, resilient animals. Our brains have made us the only technological species, but it wasn't our brains that got us here ....

It can seem to a modern human that we're a pretty weak species. Spindly little limbs, not especially fast or strong. Lots of top-level predators like lions and tigers and bears can eat us. And then there's the sharks and crocodiles ....

Why are humans so physically weak? The simple answer is we're not: we're one of the toughest, most highly adapted species on the planet and capable of a number of physical feats that other animals can't match.

Before we go on, a modern human brought up with electricity, indoor plumbing, and junk food shouldn't be compared to the grasslands hunters of 100,000 B.C. Though evidence points to us being genetically more or less identical to these pre-technological humans, they were a fair bit stronger than us, and certainly more aerobically fit. On the other hand, they only lived 35 to 45 years, often dying due to some kind of mishap. Life before history was tough.

Without his fancy technology, a human male is a medium-size predatory mammal capable of running extremely long distances and taking down prey of almost any size using a technique of harassment. Basically, we run after the animal until it's exhausted and collapses from heatstroke.

We have a number of adaptations that let us outlast lots of prey animals in this marathon-to-the-death. Antelopes, gazelles, the creatures we ultimately bred into modern cows, and many others can run faster than a human—but only over medium distances.

Crucially, many of these creatures can't sweat. They can only cool down by panting. At some point in our evolution, humans developed sweat glands, like horses—which incidentally the fittest humans can beat, too, though only over a very long course.

We also became the two-legged mammal with no tail, and the only mammal that runs upright. Even our closest relatives, the great apes, drop into a four-legged run using their front knuckles.

We have a bunch of unusual tendons and muscles in our ankles and the back of our head to stabilize us as we run. And our breathing is very clever, too. Most mammals can only breathe once per step when running (and lizards can't breathe at all when they run). We can breathe as many times as we like between steps.

All this adds up to world-record endurance. We are patient, intelligent hunters who slowly and methodically run our prey to death. Our hunts aren't as spectacular as, say, a cheetah's 60-mph sprint, but our success rate is much higher.

But in the end, it's the human brain that really gives us our edge. We remain the only species on the planet that can sit in an airliner screaming through the air at 600 mph, thinking *"Gee, I wish I could run as fast as a cheetah ...."*

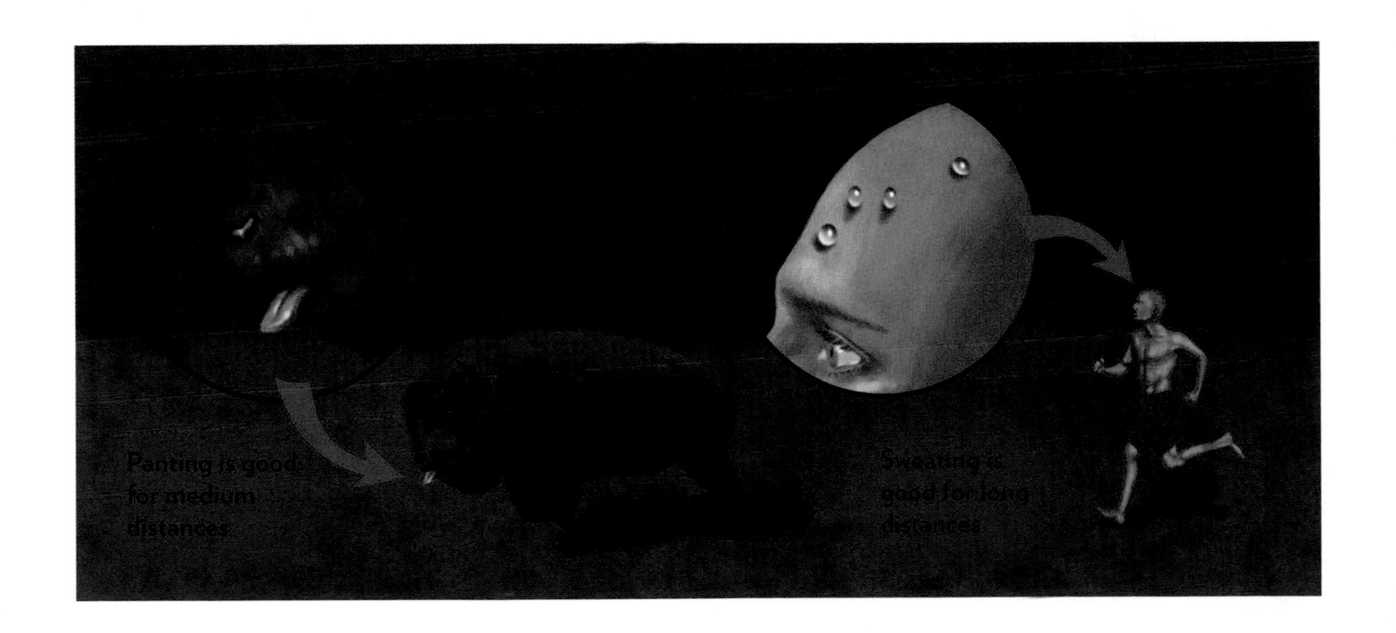

Q:

How can plants grow and regrow from one tiny patch of dirt for years?

Humans need three squares a day, but weeds can grow in a vacant lot for years and years without so much as a drop of fertilizer. How can a tiny patch of soil produce so many plants?

A:

It's amazing how little you need in life when you're literally rooted to one spot, get energy from the Sun, and can recycle your dead relatives. Yet it's all too easy to exhaust good soil ....

The plant kingdom took an evolutionary path very early on—more than a billion years ago—that allowed it to make the most of limited resources. Plants, unlike animals, developed photosynthesis—the ability to make energy by combining sunlight with carbon dioxide and water.

It's amazing how much plant you can get out of a patch of dirt. Plant an acorn, fence off an area a few yards square, wait 100 years or so, and you'll have an enormous tree weighing a hundred tons. But the soil will still be at the same level, give or take an inch. How is this possible? Where did all that tree stuff come from—enough wood to build a few wardrobes and a rec room? It didn't just come out of thin air.

Well, in fact, that's exactly where it came from—thin air and fresh water. Plants get as much as half of their entire bulk from the carbon in the carbon dioxide they take in. There's also a lot of water in a plant. Animals like humans can be 60 to 70 percent water, but many plants are as much as 90 percent water.

A big plant like a mature tree has an extensive root system that draws nutrients from deep in the ground, but what about little weeds? In an empty garden patch, weeds will grow almost as fast as you can pull them out. Why doesn't the soil run out of food?

Many plant species evolved a life cycle in which each generation would die once a year, leaving their spot on the Earth empty for their own seeds to germinate and grow. These so-called "annuals" have a ready-made source of nutrients: the corpses of their parents. And grandparents. Plant material decays and first becomes compost and then a substance called humus.

life science

Humus is organic material that can't break down (or rot) any further. It's usually dark brown because it has lots of carbon in it, and it's important for trapping water to keepthe soil humid, and also for holding on to nutrients.

Even plants that aren't annuals—these are called "perennials"—will eventually die and return their organic matter to the soil, too. In addition to this plant matter, there are billions of micro-organisms in each teaspoon of good soil, churning it and producing nutrients as part of their own life cycles. Larger animals like worms and various bugs help, too, doing the eternal job of turning large chunks of organic matter (food for them) into poop (food for plants).

Because of this system of recycling, large plant communities such as forests or grasslands can be sustained for hundreds of thousands of years. Sadly, though, the system usually ends up getting disrupted.

Before humans, disruptions included everything: natural climate change (conditions turning drier or wetter), continental drift, volcanoes, flooding, and even asteroid impacts.

These days, plant communities get destroyed mainly by us. Land clearing and poor farming practices are the main culprit—we force plants to suck the land dry of nutrients. Still, we understand how the system works, and we can take steps to stop and even reverse the damage of 100 years of industrial farming.

Q:

Is it true that most of the cells in my body aren't human?

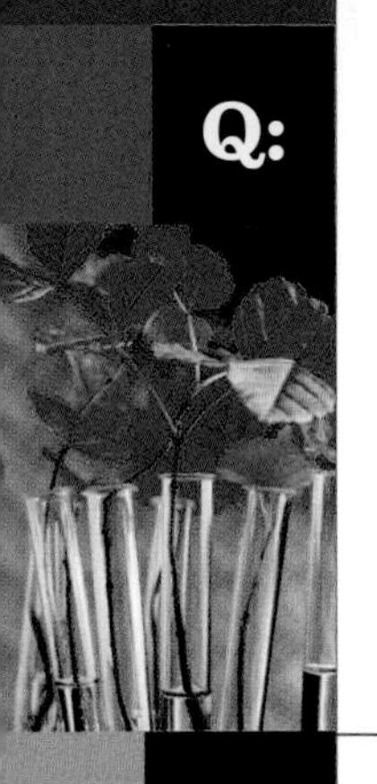

Since we need various bacteria to help with our digestion, and given that we have all these mites and other things living in our eyelashes, hair follicles, and creasy bits, are most of the cells we carry around actually not human?

A:

We do have a lot of hitchhikers, some we need and others that just ride along for the free blood. And yes—they outnumber us by nearly 10 to 1. And some can be real nasties ....

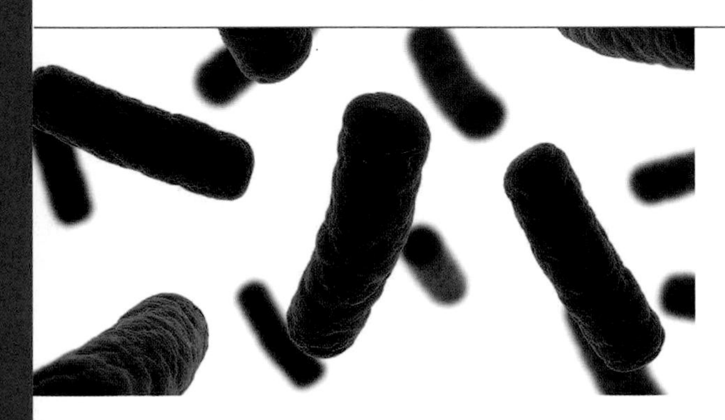

The scientifically correct answer to this depends on how you want to count cells. By weight, you are definitely mostly human, around 90 to 95 percent, depending on whether you've had a really good poop recently (sorry—but that's nature); there are a lot of bacteria in feces. But by cell *count*, only about 10 percent of the cells you haul around everywhere are your own.

Every person is a mobile ecosystem supporting a wide range of mites, fungus, bacteria, and viruses (though viruses don't have cells). There are not billions but trillions of nonhuman cells swarming over your body at every moment. It's enough to make you want to take a shower. But don't, because you might pick up even more bacteria and fungus from the bathroom.

Many of these cells live in what's called a "commensal" relationship with us. Commensal literally means "eating at the same table." These organisms—mostly mites, fungus, and bacteria—don't steal energy from us per se. They eat the stuff we cast off, or even the dirt that gets on us. Demodex mites, for instance, live in our eyelashes and eat dead skin. Others eat our sweat. Some even eat our clothes.

Then there are the organisms that live in a "symbiotic" relationship with us. This means they need us to survive, and we need them. The most important of these are the bacteria in our gut. We need these bacteria to break down our food. They get into the dead cells of the plants and animals we eat and break them open, releasing the chemicals inside. We take the chemicals we need to live, and the bacteria gets to eat the rest. It's the bacteria's own metabolism that produces methane and hydrogen sulfide gas in your intestine, which must be ... ahem ... passed.

The more you look at the nonhuman material inside the body, the weirder things get. There are possibly thousands of species of virus in us that don't appear to do anything. But there are also dormant viruses like herpes that will occasionally flare up and cause a cold sore or worse. And there are even harmful organisms that can lie in wait for our immune systems to weaken, and then pounce. Malaria is a good example—it can go dormant in the bloodstream and come back months later even if you've left malaria country.

Scientists call this community of different nonhuman organisms in the human body our "microbiome." And it's starting to look like the balance of bugs inside us can have a massive effect on our health.

We've known for some time that taking antibiotics to knock out a mild respiratory infection can also kill off huge numbers of our so-called "gut flora." This is one reason taking antibiotics can leave you feeling intestinally upset.

More recently, evidence has emerged that the mix of gut bacteria and the types we have inside us can have a huge effect on whether we become overweight. It could be why some people can eat lots of burgers and stay skinny, while others grow obese.

Creeped out by all those bacteria slithering around inside you? Think about it this way: you can't see them or feel them, and you'd get sick without them. So embrace your little friends—they might be the best friends you have.

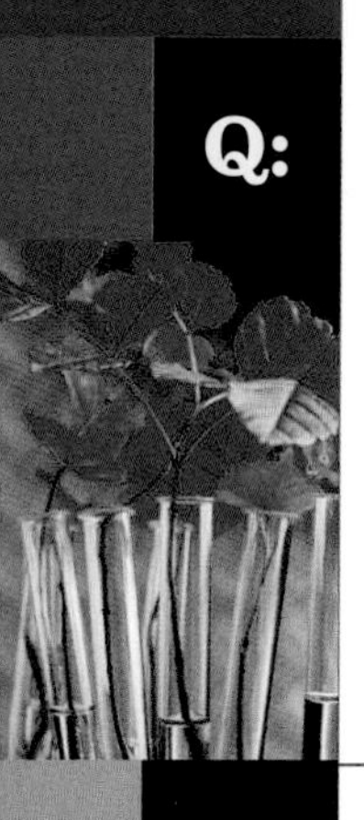

Q:

Are some birds as smart as primates?

Parrots can talk, ravens can recognize faces ... have we been underestimating bird intelligence for years?

A:

Birds are very intelligent—much smarter than we give them credit for, because their brains are quite different from ours. And they have a bunch of other physical advantages over us, too ....

Biologists have been paying a lot of attention to birds in recent decades. In the early days of serious science, birds were written off as not particularly intelligent animals, living mostly by instinct. But more recent studies suggest they may not lag as far behind us as we thought.

Scientists specializing in brains have always assumed, probably correctly, that the human brain is the most advanced thinking organ ever produced by nature. And a prominent feature of the human brain is its wrinkles: we have very wrinkled brains. Compared to other mammals, our brains are the most wrinkled. And as a general rule, the less wrinkled a mammalian brain, the less intelligent the mammal.

Birds have much smoother brains than mammals. So, naturally, scientists thought birds must therefore be pretty stupid. Unfortunately, there was quite a bit of evidence that pointed to this being wrong: birds make elaborate nests, can navigate thousands of miles, sing intricate songs, collect very specific objects with their beaks, and learn to speak human languages. There are even unproven rumors of hawks that can use fire—grabbing burning sticks and dropping them on grasslands to start fires and flush out prey animals.

Long-running studies of African Grey parrots have shown these birds are capable of not only learning hundreds of words, but actually understanding them.

A famous subject of the experiments, a parrot named Alex, could identify objects based on their color or what matter they were made of. He could identify something as "blue" or "wood." He could count objects on a tray. Amazingly, he could even "count to zero" and realize when there were no objects of a specific type visible.

Anyone who has had a large parrot as a pet can attest to their intelligence and sensitivity. The birds react with incredible empathy to the mood of their human flock-members and will get depressed or pine for absent or dead family.

It seems, then, that despite their smooth brains, birds can be as intelligent as many species of monkey and maybe even some species of great ape.

Birds are amazing creatures with unique adaptations that make them superior to mammals in some respects. The big one is obvious: they can fly. It's an astonishing adaptation that has radically changed their bodies. In exchange for giving up use of their forelimbs for manipulating the world around them, birds can instead leap into the air in what might be nature's ultimate expression of physical freedom.

Birds run a hotter blood temperature than mammals—it's like they have a permanent fever of 104° to 108°F. This is because they have a faster metabolism and the chemical reactions that go on inside their tissues need a higher temperature to operate.

Birds have unique lungs with openings at both ends (instead of just a single opening, like ours). Air flows through a bird's lungs in one direction, which means they can constantly extract oxygen and don't have to spend half their time breathing out. A complex system of air sacs allows their lungs to work like this while the bird itself still pants like a mammal.

Their hearts can pump between 400 and 1,000 times a minute, and they take as many as 450 breaths a minute. Compare that to our heart rate of 160 at a sprint and breathing speed of 30 breaths per minute!

Think about this the next time you catch a crow looking at you with an appraising eye. He's probably thinking deep thoughts.

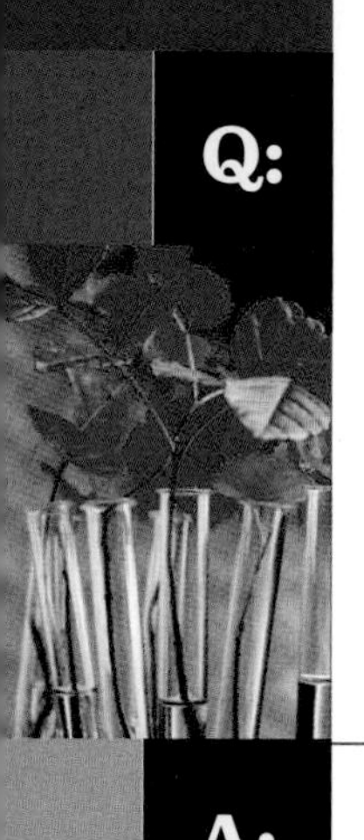

Q:

Why do some animals lay eggs?

While humans and other mammals evolved the ability to carry their babies inside them and pop them out when ready, reptiles and birds are stuck with laying eggs. Why didn't they evolve live birth?

A:

An egg gives a baby bird or lizard a miniature ocean in which to grow. And when you have cold blood or you need to fly, live birth can be more trouble than it's worth ....

When animals emerged onto land, they faced a big evolutionary challenge Newly conceived babies need to be floating in water to assemble their bodies. Sea life had already evolved simple, soft, jelly-like eggs that kept all the genetic material together while still letting seawater circulate around the embryo (fish still use these); but now life needed a way to close off the system so it worked in the air.

The first solution was the shelled egg. A land animal's egg, say a lizard's or a bird's, is a little pocket-sized ocean with just the right amount of water and raw materials for the embryo inside to grow big enough to live on land. These eggs are much more complex than you'd think. A hen's egg, for instance, has at least 15 separate parts.

The yolk of the egg provides all the raw building materials to make a bird (or lizard), and the egg white protects the yolk and the embryo and provides the water it needs for doing all those chemical reactions while building the chick.

The disadvantage of eggs is that predators can steal and eat them, or they can break. Mammals came up with a safer alternative, which was to keep the embryos inside the mother while they developed. Rather than a yolk, the embryos are supplied with nutrient-rich blood direct from the mother via an amazing structure we know as the placenta (though some mammals, the marsupials, don't have this).

The babies are kept safe inside the mother until it's time for them to be born. Some animals give birth to highly developed young—horses can walk moments after birth. Others, especially humans, need to care for their young. But even these babies are less helpless than an egg: they can cling to mum while escaping predators.

The disadvantage of having a womb and live birth is that a pregnant female is heavier and slower than normal, and she needs lots of energy to grow the baby.

Birds haven't evolved live birth because they rely so much on flight to escape predators and gather food. A bird that had to be pregnant until the chick was ready for birth would grow very heavy—maybe even too heavy to fly at all. As anyone who owns chickens knows, the hens are "pregnant" with their eggs for a very short time—only 26 hours.

What about lizards? Well, reptiles are cold blooded. They don't need as much energy as a hot-blooded animal like a bird or mammal, but the trade-off is that they can't maintain an internal temperature that's right to incubate their eggs. Almost all reptiles bury their eggs, relying on the insulating properties of the Earth to keep them the right temperature.

Of course, since evolution likes to mess with us, there *are* groups of lizards that give birth to live young; but really the mother is just incubating and hatching her eggs inside her body—she doesn't have a womb like a mammal. These species tend to live in warmer climates, too.

Egg laying is not more "primitive" than live birth, it's just a different solution to a common problem: how to grow a baby when you don't live in the ocean.

Born in the ocean

Mother lives on land

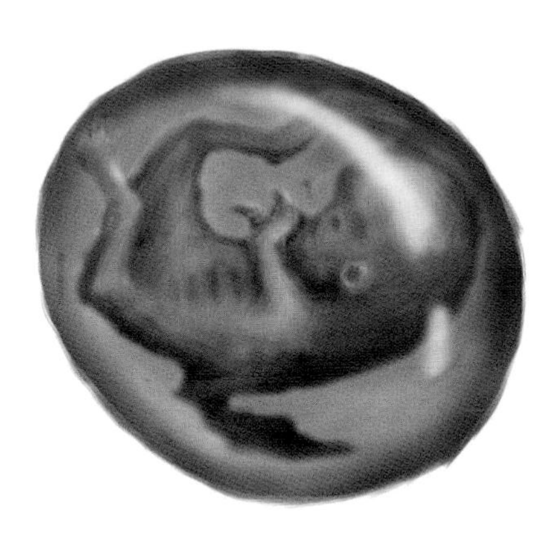

Mother must travel or flee

Q:

Why do all large animals have four limbs?

When we look in the ocean, we see an amazing variety of life forms: finned fish, tentacled squid, ten-legged lobsters. But on land, every large animal has two arms and two legs. Why?

A:

The four main types of land animals—amphibians, reptiles, birds, and mammals—all evolved from a single group of amphibians in the Devonian age, over 400 million years ago. These were four-legged, too—and if it ain't broke, evolution don't fix it ....

The way every vertebrate—that is, an animal with a backbone—has four limbs is good evidence for evolution and a common ancestor. Even snakes and whales have genes that "switch off" their legs and stop them growing. But there are no naturally three-, five-, or six-legged vertebrates.

All land vertebrates are part of the superclass Tetrapoda, which is Greek for "four-footed." And the simple answer is that we all have four limbs because we have a common amphibious ancestor who also had four limbs.

Evolution operates on the principle of keeping what already works efficiently. When an animal finds a perfect niche, it can maintain the same basic body shape and physical abilities for millions of years. Crocodiles and alligators are a great example—the "idea" of a crocodile (scales, big teeth, lives in a swamp) is over 80 million years old. They are really good hunters, and there's just no need for them to evolve.

Vertebrate life came up onto land during the Devonian period. Why? One main reason is because atmospheric oxygen levels started to climb. Oxygen in the air back then was only 15 percent—it's 20.95 percent now—but even at that level, there's much more oxygen per breath in air than in water.

Fossils show that fish were evolving lungs before they even thought about walking on land. Back then, a lung was an air sac that could extract oxygen from the air. Fish, living in shallow lakes and rivers, would gulp air into the primitive lung and process it while still breathing water through gills at the same time. This fish lung has since evolved into the swim bladder that many fish have, which keeps them from sinking to the bottom.

Meanwhile, evolution had to come up with a way to let fish and amphibians (air-breathing creatures that still need water, like today's frogs and newts) navigate through shallow streams and brooks that were choked with fallen branches and leaves from another newfangled kind of land life: plants.

The solution? The amphibian's four fins became four flippers, which became four legs, which eventually became strong enough to allow these creatures to support their own weight in air.

Lungs became more efficient, gills were discarded, skin got hard and stopped drying out, and many other adaptations were tried. Some of these tetrapods even evolved flight, turning their forelimbs into amazing wings.

Even though millions of species evolved with lots of different physical abilities, the sheer mechanical efficiency of four limbs—support for four corners of a body, and the ability to lift two off the ground at once and still not fall over—was never bettered.

So what's the deal with insects, spiders, centipedes, and so on? Since they are so much smaller than us, individual bugs need less total energy to grow and live. This gives adaptations a better chance to survive. Adding an extra pair of legs to a bug doesn't add that much more total energy. And because arthropods (the collective term for all bugs and creepy-crawlies) are mechanically simpler than vertebrates, it's easier for changes to happen. Evolving a new leg doesn't require many other parts to change as well.

The four main types of land animals, amphibians, reptiles, birds, and mammals all have four limbs

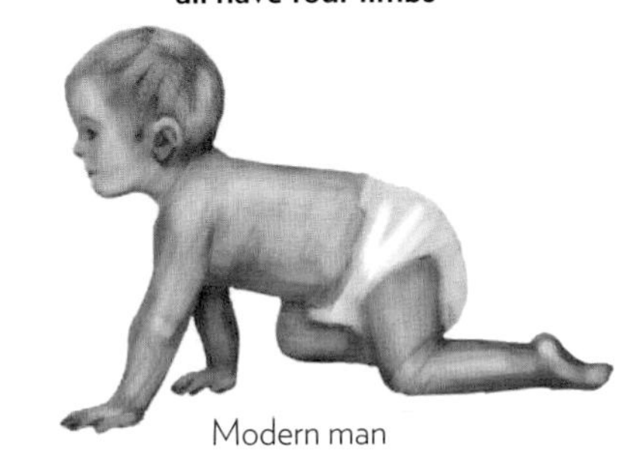

Modern man

Allosaurus

Panderichthys

Coelacanth

Q:

How do insects and spiders breathe, and why can't I see them breathing?

Bugs and other creepy-crawlies breathe oxygen just like us, but even if you look closely at a bug you can't see its little chest going up and down or its mouth panting. How do they get their oxygen?

A:

Life is simpler when you're really tiny. Mammals, birds, and reptiles need to pump air into their lungs. Bugs just let air flow through holes in their bodies and into their blood. Except it's not really blood ....

Amphibians, reptiles, birds, fish, mammals—we all breathe oxygen from air or water, and we put that oxygen into our blood. The blood is then circulated around our bodies and the oxygen fed to our cells to create energy.

Insects have a system that's almost entirely different. Instead of putting oxygen into blood and transporting it through veins, insects have a different system of tubes branching through their bodies called tracheae, capped with holes to the outside world called spiracles. This internal plumbing delivers oxygen directly to the tissues.

Insect "blood" is a fluid called hemolymph. It's not true blood, but rather just sort of bathes the insect's internal organs in a nutrient-rich soup and cells take what they need. This means insects don't need lungs. They are small enough that air will be drawn into their bodies by air pressure alone.

Early understanding of insect anatomy led scientists to believe the oxygen just permeated an insect's tubes with no physical help. Now, after decades of research, we can see that some insects have tiny pumps to move air through the tracheae, and they can open and close their spiracles with muscles to stop water escaping through the tubes and drying them out.

Insects are part of a group called the arthropods, which means "jointed leg" in Greek, and includes spiders, centipedes, crabs, lobsters, and more. Basically anything with a hard outer shell, lots of jointed legs, and no backbone is an arthropod.

Many arthropods use the same spiracle-tracheae breathing system as insects. The ones that live in the ocean use gills. But there's another system, too, used by spiders and scorpions, called a "book lung."

In a book lung, layers of tissue resembling the pages of a folded book are arranged with gaps of air in between. The tissue is full of hemolymph, and oxygen seeps in. Spiders don't even need to move their book lungs to get oxygen.

Not all spiders have book lungs; some have one pair, and some scorpions have four pairs. This is the thing with arthropods: there are so many different kinds with different ways of doing things that it's very hard to come up with a simple answer for how "all" insects breathe.

Grasshoppers, for instance, have several tube-shaped hearts along the sides of their body. Other insects only have one heart. Spider hearts are very simple, but most insects have a number of chambers in their hearts, all in a row.

But they all have this fascinating "open circulatory system" that supplies organs with nutrient-rich fluid and oxygen. It's why bugs go splat when you swat or stomp them. Without an internal skeleton, veins, or arteries, the inside of a healthy uninjured insect is more like a soup with lumps and stringy bits.

The advantage is that without complex skeletons and blood vessels, arthropods have been able to evolve into a huge range of amazing shapes and sizes. The biggest ever arthropods were the sea-dwelling, scorpion—like eurypterids and a giant millipede-like critter called arthropleura. Both could grow to eight and a half feet (2.6m)!

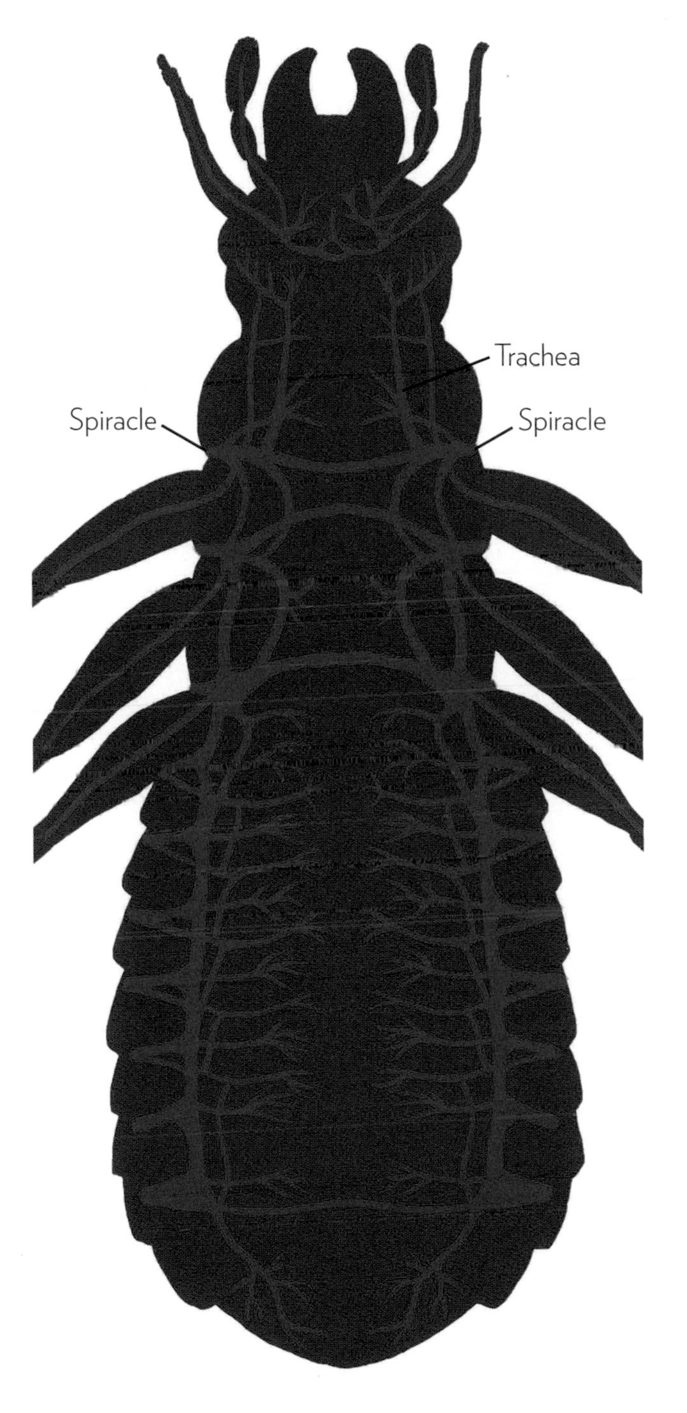

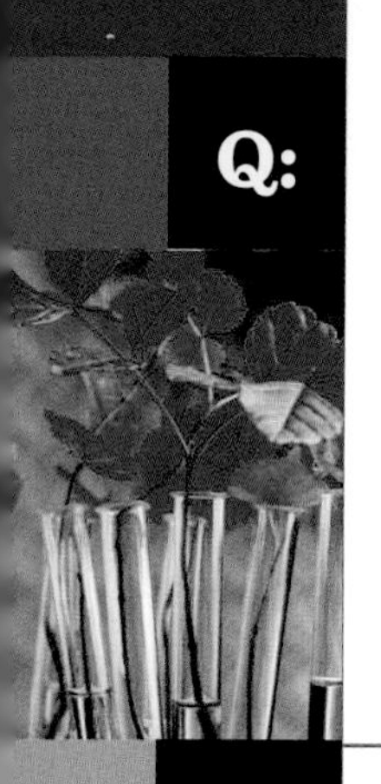

Q:

Did dinosaurs have warm or cold blood, and how would we tell anyway?

Birds have warm blood, and birds are descended from dinosaurs. So does that mean dinosaurs had warm blood, too? How can we tell just from fossils?

A:

There is evidence for and against dinosaurs having warm blood. On the one hand, the way their bones grew suggests warm blood. But their skulls are missing structures that all warm-blooded animals have. So, warm or cold? The real answer could be neither ....

Trying to get an accurate picture of what dinosaurs were really like from the fossil record alone is like trying to do a jigsaw puzzle you found buried in your great-grandparents' yard. There's no picture of what the puzzle should be, lots of pieces are missing ... and you're not even sure it really *is* a jigsaw puzzle. It could be a broken jug.

Over the century or so we've been studying dinosaur fossils with real scientific rigor and sophisticated instruments, we've come to learn a lot. We know that dinosaurs weren't all slow, heavy reptilian creatures; some were quick-witted, fast-moving animals more like today's birds. Giant, deadly birds. Speaking of birds, we've also figured out that birds are descended from one group of dinosaurs called the theropods. A bird is more closely related to a Tyrannosaurus rex than a lizard.

While birds have a number of similarities to reptiles, even if you leave aside their flight they have one more major difference. Birds are warm-blooded, and most species have even hotter blood than humans.

Does this mean that the theropod dinosaurs were also warm-blooded, or did birds suddenly evolve this major metabolic difference later? The answer may not be cut and dry.

Some paleontologists believe at least some dinosaurs must have had self-heating blood because many were too big to have survived otherwise. Warm-blooded animals pump blood faster and harder than cold-blooded ones, and this would allow animals like the Brachiosaurus to have really long necks.

But on the other hand, some dinosaurs were too big to have a modern hot-blooded system. A massive sauropod weighing over 100 tons would probably overheat and die.

Dinosaurs are also missing important structures in their skull called "nasal conchae." These are a system of curled bone shelves in the nose that divert air over a very large surface area and work as a natural air-conditioner. Since warm-blooded mammals breathe seven times faster than cold-blooded reptiles, we'd risk dehydration if we let our exhaled air carry off too much water. The nasal conchae trap water and return it to the body. But dinosaurs don't have this feature, even though birds do—and this is evidence against them being warm-blooded.

Or is it? Maybe their nasal conchae were made of cartilage or some other material that doesn't fossilize. Maybe a thousand other possibilities ....

The problem is that the last dinosaurs died out 65 million years ago, and evolution has come up with a lot of new stuff since then. It's incredibly difficult to figure out if dinosaurs break the modern rules of cold- or warm-blooded animals ... or if maybe they were something else altogether.

Some scientists have suggested dinosaurs didn't heat their blood via chemical reactions like we do, but warmed up in the Sun like a reptile—yet unlike a reptile it took dinosaurs a long time to cool down again. They may have had clever insulation that maintained a high body temperature.

If dinosaurs did have this not-hot, not-cold metabolism, it might explain why some fossils show evidence of warm-blooded bone growth or a body shape that implies fast movement and lots of activity, while other dinosaurs are of a size or shape that implies cold-blood. The debate rages on!

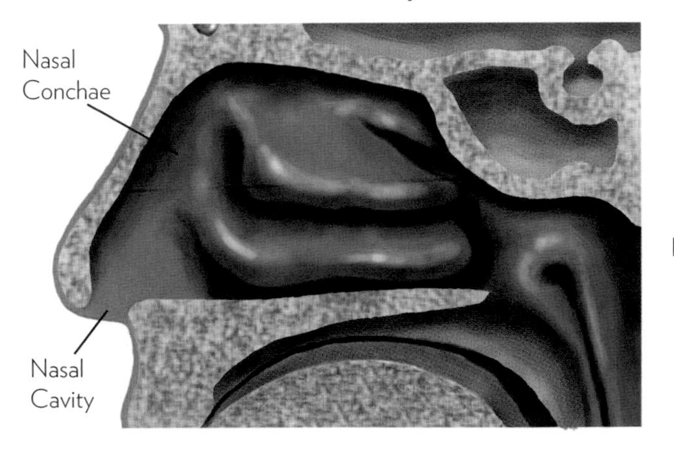

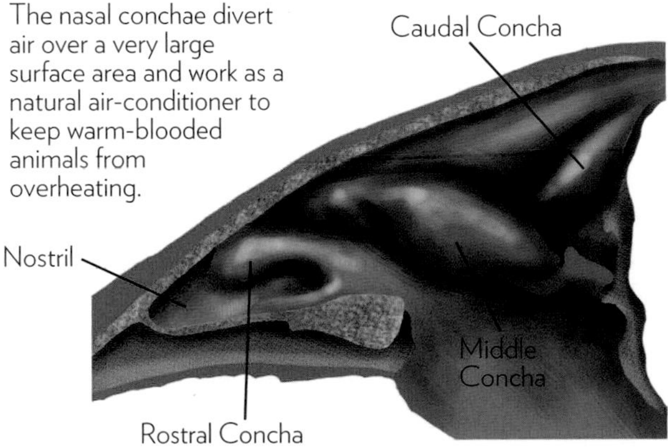

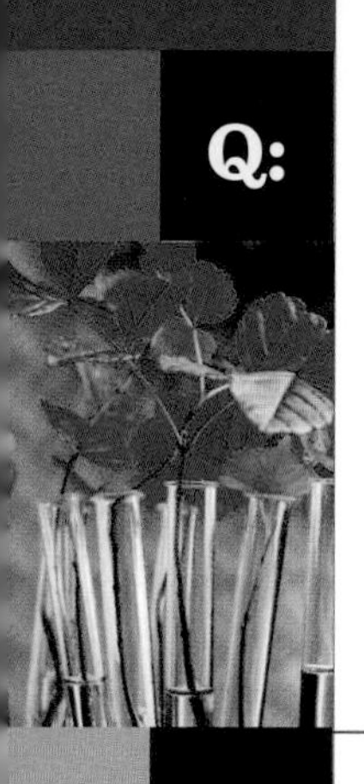

Q:

How do we heat our blood, and why is it a particular temperature?

Mammals and birds have an internal body temperature that's precisely regulated and hotter than the air (on average). What's the benefit of warm blood, and how do we heat it up in the first place?

A:

Chemical reactions in our cells produce heat, and our bodies can very precisely maintain our internal temperature. Hot blood makes us very energetic, but there's another good reason for it ....

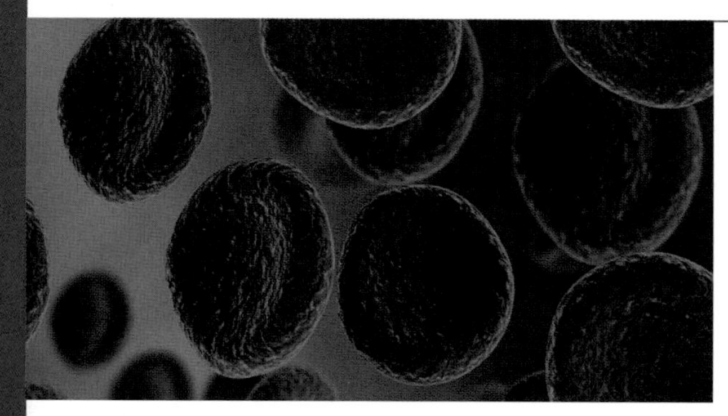

Birds and mammals have an important adaptation that reptiles, fish, insects, and plants don't: we can control our internal temperature very precisely.

As every parent with a sick child knows, the ideal human body temperature is 98.6°F, though individual people can vary by nearly a degree and still be healthy. Birds are usually a bit hotter than us—normal is about 104°F (a deadly fever for us!)—because they need access to lots of quick energy for flight and so have faster cellular processes.

Scientists call these processes the *metabolism*—another Greek word that simply means "change."

Warm-blooded animals have very precise control over their metabolism. The nervous system continually monitors body temperature, and if it gets too cold will use up some sugar—which is like throwing more wood on the fire. Chemical reactions release heat, and the heat warms up our body.

So-called cold-blooded animals like reptiles and fish actually use very similar chemical reactions to us. All animals have these metabolic processes in their cells, where we break down food and mix it with oxygen to create energy we can use.

This process is pretty inefficient. Up to 60 percent of the energy is lost as simple waste heat. So all things being equal, after eating, say, a mouse, a lizard and a weasel would potentially generate the same amount of heat from digesting and metabolizing the same-size mouse. The difference is that the lizard would just let all the heat leak out through its body.

Warm-blooded animals have insulation such as feathers, fur, or blubber to trap the heat, and their metabolism generates more heat in the first place by running the chemical reactions faster. We're also very good at increasing the rate we use our chemical fuel when we exercise. All this combines to keep our bodies hot.

You might have heard of a substance called "ATP"—it's often mentioned in conjunction with anti-aging products or health supplements. It stands for adenosine triphosphate, and it's a chemical compound packed with energy that we use to power our cells.

Our nervous system monitors the amount of ATP in our tissues. One way we deal with getting too cold is to shiver. This rapid muscle movement uses up all our ATP really fast. Our body then makes more ATP, and in making it generates lots of heat.

We can also increase our internal temperature to kill off bacteria and viruses. We call this a fever and usually think of it as a bad thing, but fevers save us from infections running amok (and we cool down again by sweating or panting).

In fact, this feverish or "febrile" aspect of warm-bloodedness could be why we evolved it in the first place: it protects us not just from bacteria and viruses, but against getting infected by fungus. Reptiles often suffer from terrible fungal infections, but we rarely get anything worse than thrush (though this can kill small children).

This is why our body temperature is 98.6°F. It's the temperature that's high enough to kill off most infectious fungus, but it's not so high that we'd need to spend the whole day eating and building up our energy. As always, evolution picks the most efficient way!

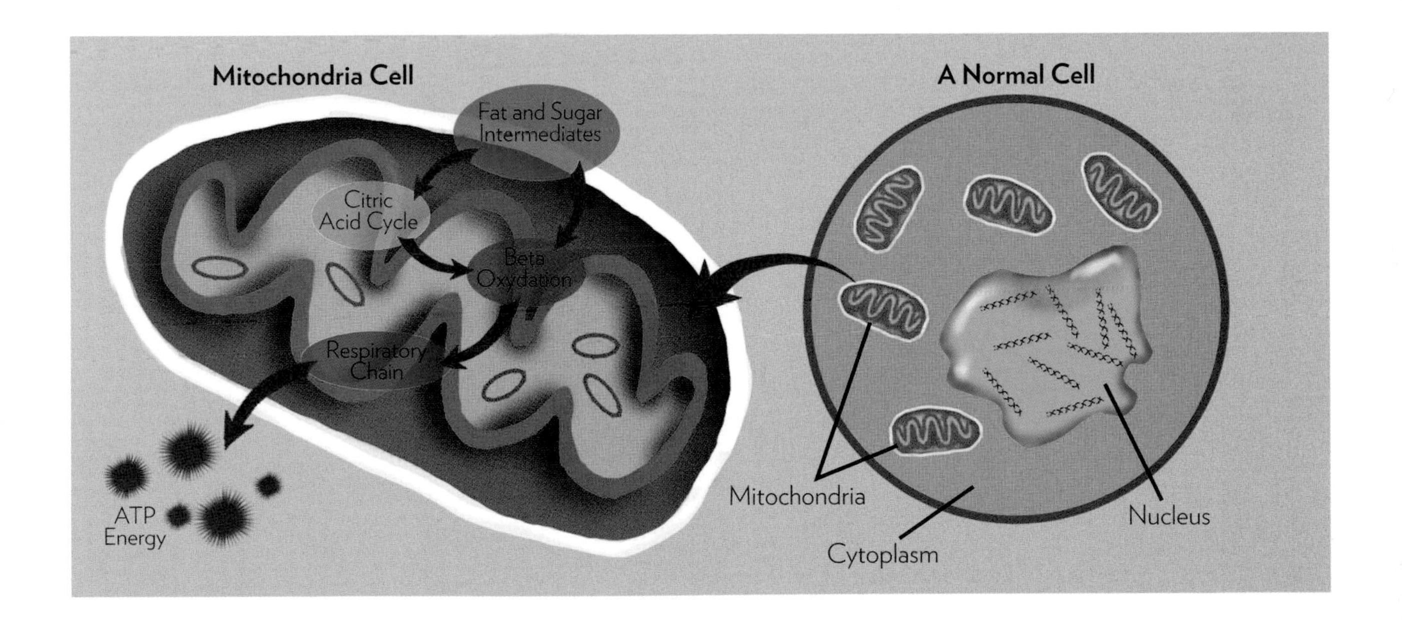

Q:

Why can I heal a deep gash in my arm, but can't regrow a lost tooth or fingertip?

Healing is an amazing ability: if we cut ourselves, our skin will close up and seal itself. Scrapes and grazes regrow skin. But it only goes so far. Why can we heal a cut but not regrow an adult tooth or a missing finger?

A:

Some amphibians can regrow lost limbs, but it takes a really long time: more than a year. We need to heal fast so we can survive. And there are worse things than not being able to regrow a limb ....

A human's healing ability is pretty poor compared to some other members of the animal kingdom. Newts and salamanders can regrow entire limbs—bones, nerves, muscles, and all. Many lizards can regrow their tails. And more primitive creatures like starfish and some flatworms can regenerate huge portions of their bodies, good as new.

When a human gets badly injured, losing a big chunk of muscle or a finger, the body responds by using stem cells to generate new skin and cover the wound. Eventually we grow a fibrous material to keep the wound closed, and if the injury was big enough you'll be able to see this as a scar.

Salamanders, on the other hand, respond to an injury quite differently. If one of these amphibious, lizard-like creatures gets a limb bitten off by a predator, stem cells cover the wound, but instead of forming a scarred stump, they form a structure called a "wound epidermis."

In a process still not fully understood, stem cells swarm and start to build a tiny, almost embryonic version of the missing limb. They'll even take apart surviving healthy tissue so they can start with a clean foundation. After a few weeks, the salamander will be sporting a tiny but complete new version of its old leg. This new limb will grow slowly over time. Over a very *long* time. A small salamander can take more than a year to fully regenerate a leg, while a full-size one can take more than 10 years! At this rate a human limb weighing many pounds would take decades to regrow.

The thing about amphibians is they live slowly. When injured, they can hide away and sort of shut down their system, using very little energy, and dedicate what stores they do have to healing. Warm-blooded animals like us need to eat constantly; we just don't have the time to sit around waiting for a finger or hand to regrow. We need to heal fast—and that means giving up the ability to regrow lost limbs.

That doesn't really explain why we can't regrow a kidney if we have one removed (after all, it's safe in our body and we can live just fine with one kidney in the meantime). And it doesn't explain why our liver *does* regrow if we get a tumor removed from it. The liver is the only organ that can do this, though technically it just extends the old tissue that's left over from surgery; it doesn't regenerate an identical new liver. Still, we're sure amputees wouldn't mind if their missing legs regrew as different legs.

The answer to this puzzle might be the same answer we've given for a few other mysteries (such as why we age and die). And that answer is cancer. If we could regrow our fingers, it might massively boost our chances of getting cancer.

Medical researchers are hard at work trying to unlock the secrets of tissue regeneration. But it may take so long that by the time we figure this out, we'll be able to regrow organs and limbs in vats. Why walk around with a miniature arm for 10 years when you can just order one up and bolt it right on?

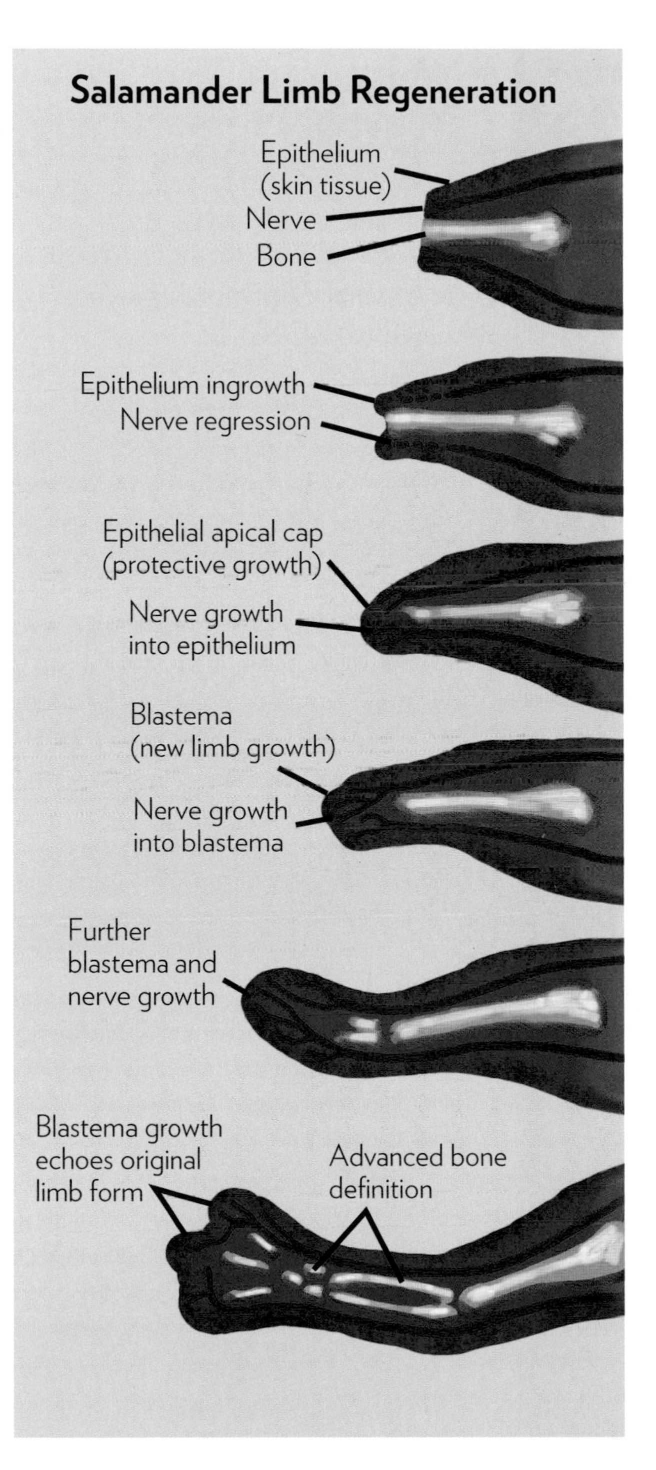

Q:

Why can't I breathe water even though a fish can (sort of) breathe air?

When fish get stranded on the beach, they sometimes lie there gasping for 15 minutes or even longer. But humans drown in three minutes. Why can fish breathe air a little bit, but we can't breathe water at all?

A:

Both fish and humans breathe oxygen gas, but for fish the gas is dissolved in water. Water holds 20 times less oxygen than air, so a fish's gills—though not designed for it—can extract some oxygen from air. Humans may one day be able to breathe liquid, though ....

Part of the reason life evolved to live on land (apart from all the free real estate) is that the atmosphere contains much higher concentrations of oxygen than seawater—up to 30 times as much, depending on conditions.

Land animals can be much more energetic than sea life, because we can suck in so much oxygen for our fast metabolisms. Our lungs have evolved from gills to take in air.

But at the final stage of oxygen extraction, we actually dissolve the gas into a liquid (our blood). The only big difference between us and fish is that the fish don't need a clever system to dissolve the oxygen into water. They just breathe the water.

But because seawater has such a low concentration of oxygen, a fish's gills need an absolutely massive surface area. They are very complex and ornate, with many branching structures. Our lungs, on the other hand, don't need as much surface area because there's so much oxygen in the air.

This is why taking a fish out of water isn't as immediately fatal as dunking a human in the deep end. The gills can extract oxygen out of the air no problem—except there actually *is* a problem.

Gills do not support themselves, they rely on buoyancy in the water to stay open and spread out and able to catch the most oxygen. When you pull a fish out of the sea, its gills collapse against themselves. There's enough gill still working to extract some oxygen, but not enough to keep the fish alive. They suffocate, quite slowly.

Human lungs don't have as much internal surface area, and they are designed to pump gas in and out. And since water is so much thicker than air, we can't pump it in and out of our lungs fast enough. And since there's so little oxygen in water compared to what we're used to, we run out of oxygen much faster and drown in just a few minutes.

Actually fish can drown, too—if they get stuck or held in a way that means they can't open and close their gills, they run out of oxygen. Or they might stray into a so-called "anoxic" area of the ocean where there's not much dissolved oxygen. They'd drown quickly there, too.

Having lungs full of a gas can be a disadvantage for humans when we want to mess around in areas of very high or low pressure. Divers are limited in how deep they can go because they need gas for their lungs.

But there is a liquid humans can (theoretically) breathe. It's a type of fluorocarbon that's very rich in oxygen. As well as helping divers, it could be very useful for patients with certain respiratory diseases—especially children. Doctors could fill the whole of the lung with fluid, or just the bottom 40 percent of the lungs. It could be a real boon for premature babies whose lungs should still be full of amniotic fluid.

Astronauts might use liquid breathing one day, too. It would allow them to accelerate at faster speeds without getting injured by the gas compressing in their lungs, because these liquids do not compress.

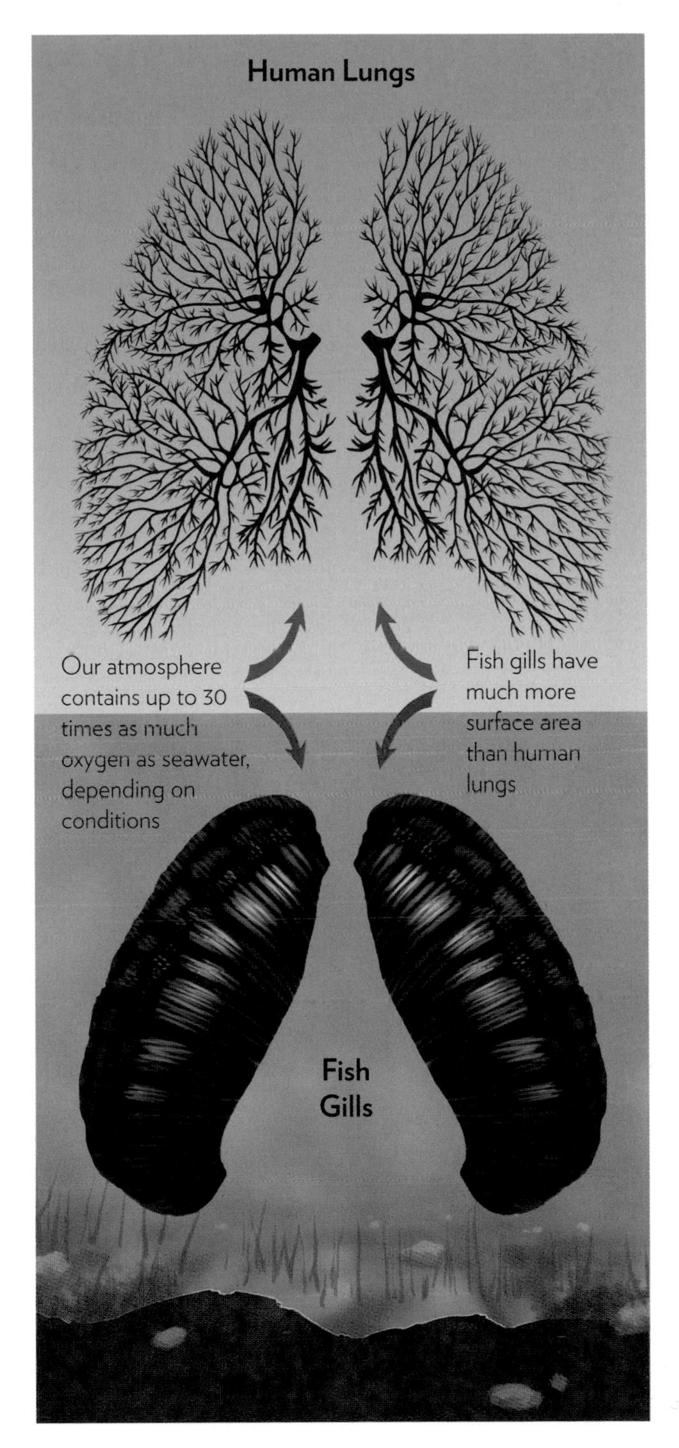

chemistry

Everything we do, every move we make, is only possible because of chemistry ....

Chemists have a saying: what in the world *isn't* chemistry? In a universe made of atoms and molecules, nothing happens without some kind of chemical interaction being involved.

From starting a fire to simply lifting your arm, chemistry makes it happen. The way atoms join up into molecules and then move energy between other molecules is what makes life possible.

Chemistry cooks our food, smelts our steel, grows our crops, and propels our cars down the road—which was also built thanks to chemistry.

Chemistry brought us into the world, and chemistry will take us out of it, too, strapped to enormous rockets. With command of chemistry, we can conquer the galaxy.

It also helps us understand the risks and challenges that will face us in the centuries ahead. Climate change, pollution, cancer, obesity, and more are all, in some way, problems of chemistry.

Q:

How many elements are there really?

We're taught about the Periodic Table of Elements in school, but scientists keep making new elements in the lab. So how many elements are there really?

A:

There are 98 naturally occurring elements, but we're able to make at least 20 more. But the real total is still up for debate ....

All physical matter on Earth is made up of a mix of just 98 different kinds of atoms. These atoms are called "elements," and we can tell them apart by their physical properties.

In the everyday world, it's easy to see that gold and silver are different elements because they have different color and weight.

Scientists write down all the elements in an oddly shaped grid based on their chemical properties, called the "Periodic Table of Elements." If you haven't seen one hanging on the wall in science class, well, you must have been asleep!

Each entry on the table is a different element, and each element has its own unique number. Hydrogen is 1, oxygen is 8, and so on.

We've had the philosophical idea of atoms for thousands of years, but the science of chemistry only discovered the atom for sure in the late eighteenth century.

Using chemical reactions, scientists—mostly in England and France—were able to break substances down into what they called "elements." Over the next couple of hundred years, chemists figured out what exactly made these elements different.

Each element is a type of atom made up of three different particles: protons, neutrons, and electrons. Protons and neutrons clump together in the nucleus of the atom, while electrons orbit the nucleus.

Atoms can have different numbers of protons in their nucleus, and this is what makes them different elements. Oxygen, for instance, has eight protons. Gold has 79. And hydrogen, the most basic element, has just one proton. It's the number of protons that gives the element its unique number and place on the Periodic Table.

Simply changing the number of protons makes a massive difference to the physical properties of an element. Oxygen, by itself, is a gas that helps our cells make energy. Gold is a heavy, shiny metal. But the only difference is 71 extra protons!

Actually, gold also has extra neutrons and electrons to match its 79 protons, which is what makes it a "stable" element that can clump together into gigantic molecules we can use to make wedding rings and stereo connectors.

There are 80 elements that occur in nature that are stable. These atoms won't change or "decay" in natural processes. The heaviest of these—the one with the most protons—is lead. The next 18 naturally occurring elements are all "unstable." That means they're radioactive and over time will lose protons until they turn into a more stable element—again, often lead! Famous radioactive elements include uranium, plutonium, and thorium.

As chemists and later particle physicists gained a better understanding of how atoms work, they realized it should be possible to mush protons, neutrons, and electrons together to actually make new elements. So over the last 70 to 80 years, we've added another 20 elements to the Periodic Table. They have weird science-lab names like "technetium" and "californium." Many of these are very radioactive and only last for a few seconds before decaying into a natural element.

Synthetic elements help us understand how atoms work and are important tools for advancing the science of chemistry. Our ability to make synthetic elements means the total number of possible elements isn't yet known for sure. Current theory suggests the max might be somewhere around 135. It just depends on how many we can make!

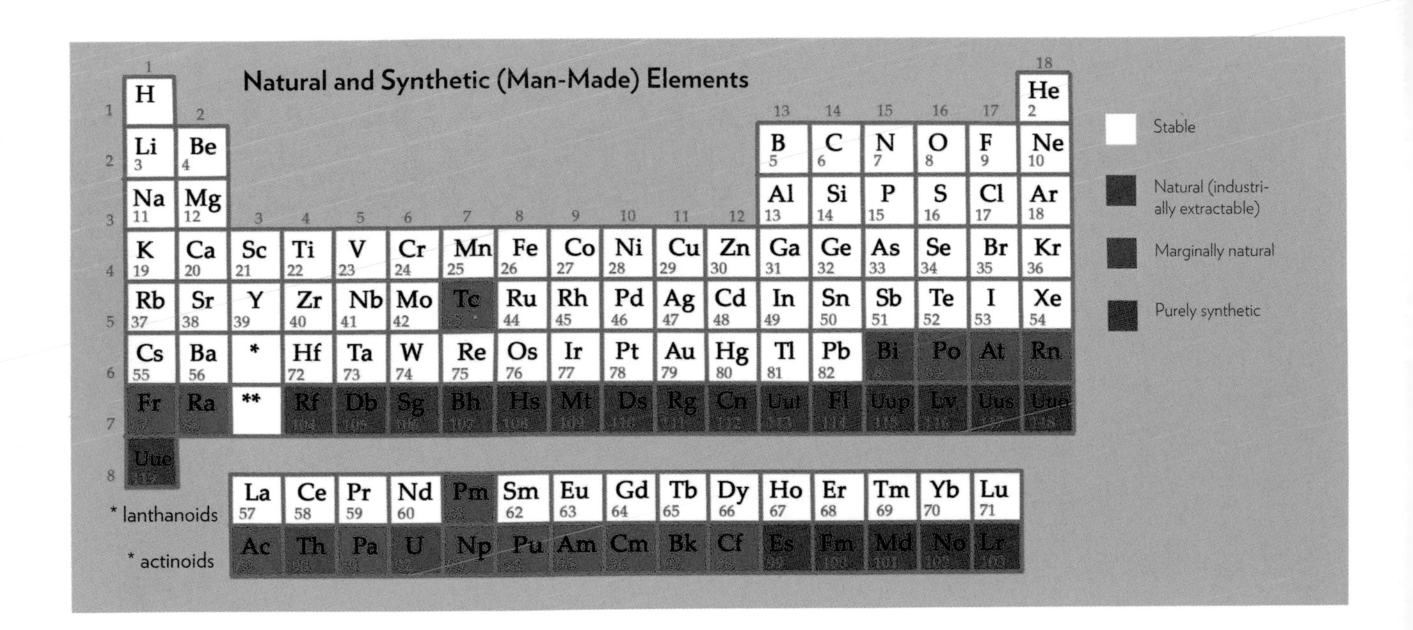

Q:

Why are some elements radioactive?

Most of the elements are stable and harmless, but some of them are radioactive and deadly to even stand near. What makes them radioactive, and why is this dangerous?

A:

Atoms have a nucleus, and if this nucleus isn't perfectly balanced and stable, the atom will shoot out particles until stability is reached. This is the essence of radioactivity.

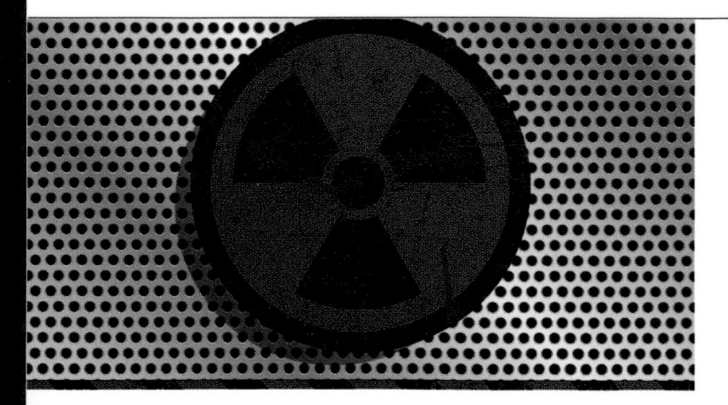

Normal physical matter is made up of atoms. Each atom in turn is made up of protons and neutrons in a nucleus, orbited by a bunch of even tinier particles called electrons.

A big object like the Earth holds its shape due to gravity. But atoms are so tiny that by themselves they can't rely on gravity to stick together. Instead it's all about electric charge—protons have a positive charge, and electrons have a negative charge. So protons and electrons are attracted to each other.

The problem, for an atom, is that protons repel each other. So if you bunch up a whole lot of protons in a nucleus, they want to fly apart. This is where the neutrons come in. They have no electric charge, but they do stick to the protons using another fundamental force called the "nuclear force." This is strong enough to overcome electric charge.

Two forces, one pushing and one pulling. So an atom needs just enough neutrons to keep its nucleus stable.

Most of the elements are stable. But some, such as uranium and plutonium, are not. The ratio of neutrons to protons in their atomic nucleus isn't "just right," and so something has to give.

Nature wants its atoms to be stable, so the atom actually throws out some of its particles to try to reach that stability.

This throwing out—or emission—of particles is called radioactivity. There are different kinds of radioactivity, depending on what kind of particles get thrown out of the atom.

A radioactive substance might emit a so-called "alpha particle," which is made up of two protons and two neutrons. Or it might change one of its protons into a neutron and shoot out an electron. This is called a "beta particle."

The result of both kinds of decay is that the total number of protons and neutrons in the nucleus of the atom changes. Over time, it will end up decaying to an arrangement that's stable.

There are other kinds of radiation, too, including x-rays and gamma rays, all depending on the type of radioactive element you're dealing with.

The real danger to humans from radioactivity is these emitted particles. That's because the particles are so-called "ionizing radiation"—they carry an electric charge. Why is that bad? Because particles with an electric charge can react with other elements—including atoms and molecules in your body.

The alpha particle, with two protons, carries a positive charge, while the beta particle carries a negative charge. And they get fired out of the radioactive element at high speed. Getting hit by one of these particles is like getting hit by a car. The "kinetic energy" of the particle gives it the power to smash apart chemical structures in your body.

You get hit by radioactive particles all day every day, but usually in very small numbers that don't do much harm. But if you get hit by millions of these at once, at a microscopic level your tissues can end up looking like Swiss cheese.

A moderate dose will damage your DNA and give you cancer. That's bad, but at least you can get treatment. Really high doses of radiation, like you'd get from standing next to an unshielded nuclear reactor, are so powerful they can burn your skin, destroy your blood, and kill you within minutes.

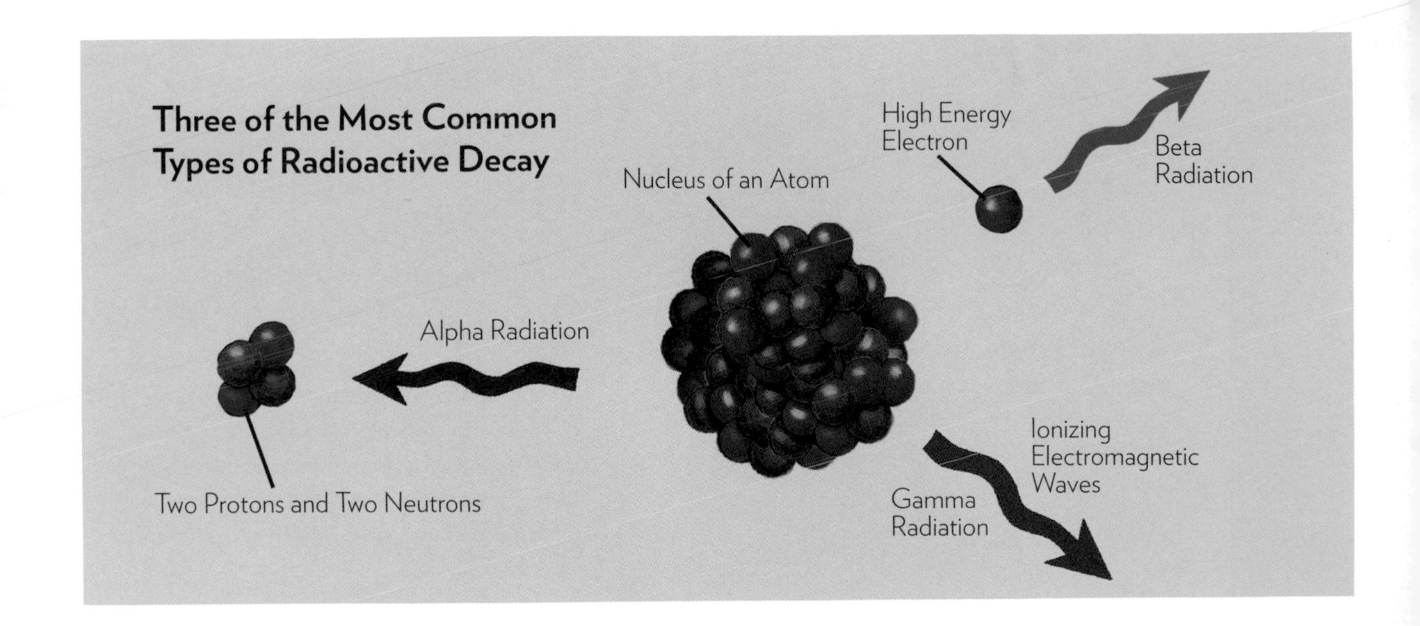

Q:

Why does lead protect me from radiation?

Radiation is made up of tiny waves and particles that enter the body and cause cellular damage. But some materials, including lead, block radiation. How is this possible?

A:

Lead is the densest, heaviest nonradioactive element in nature, so it blocks a lot of radiation. But it turns out radiation isn't that hard to shield against after all ....

Radiation comes in a number of different forms, including particles made up of neutrons, protons, and electrons, and high-energy waves like gamma rays and x-rays. Even ultraviolet light from the Sun is a form of radiation (and sunburn is a form of radiation poisoning!).

What all these types have in common, though, is they are emitted in a stream that passes through your body and causes damage to your cells and tissues.

To work safely with radioactive samples, scientists use various different kinds of shielding. Nuclear reactors are also shielded. Shielding is talked about in terms of its "halving thickness." This is how thick of a shield you need to absorb or deflect half of the radiation coming from a given sample.

Almost everything deflects or absorbs radiation to some extent, but some substances do it much faster. Lead is one of the most effective radiation blockers in nature, with a halving thickness of just 0.4 of an inch (1.25cm).

Steel needs an inch, regular concrete needs 2.4 inches (6cm), water needs 7.2 inches (18cm), and open air needs about 16 yards (14.5m).

Remember that thickness only blocks half the radiation. You need to double-up to stop the majority of radioactive particles coming through the shield. But lead shielding can be much thinner and more practical than steel or concrete shields.

Lead works so well because it's very dense. With 82 protons in its nucleus, lead is the densest non-radioactive element that occurs naturally. Interestingly enough, uranium itself is an even better shield against radioactivity—up to five times better at stopping gamma rays. And because it only gives off alpha particles, we can coat it in a thinner secondary shield to make it even safer. But it is very expensive, so lead is more common.

Alpha particles, with both neutrons and protons in them, are actually pretty easy to block. You can stop most of them with a sheet of paper. Beta particles, made up of an electron, penetrate deeper. And while they can be stopped by a thin sheet of, say, aluminum, this can produce x-rays as a by-product—which are still dangerous.

If the radiation is a stream of neutrons, that can be doubly dangerous because those neutrons can hit the shield and make some of the atoms in it radioactive themselves.

Because of its density and stability, lead is immune to a lot of these effects, and so extremely good at blocking all forms of radiation. It's heavy, though, and for really big sources of radioactivity—like nuclear reactors—it gets used in conjunction with special super-dense concrete and even plain water in a clever multilayered shield.

Heading out to stock up on lead plates now? Don't worry, in your day-to-day life, you're pretty well shielded from radiation. The Earth's magnetic field and atmosphere block nearly all the dangerous radiation from the Sun—and you can block ultraviolent rays with a thin smear of sunscreen.

Anyway, in one of nature's cruel ironies, lead itself is very poisonous to humans in other ways, so we need to keep its use to a minimum. That's why we don't line our homes with lead.

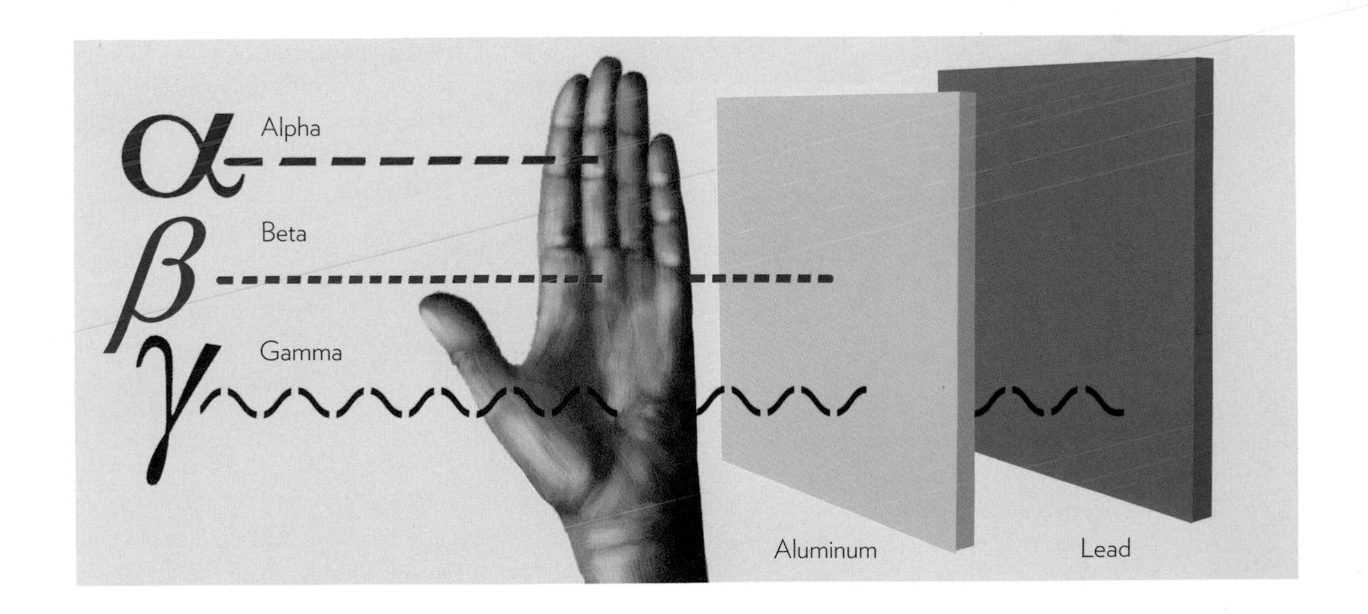

Q:

What keeps molecules stuck together?

Everything is made up of atoms, but atoms also stick together in different combinations to make molecules. How does this work?

A:

An electric charge keeps atoms stuck together via their electrons. There are a couple different ways this works, and this has a big effect on how a molecule looks.

We can't see individual atoms without using a special kind of instrument called a scanning electron microscope, but we can see individual molecules.

While some are still very tiny, other molecules are huge. Most pure metals—for instance, gold, aluminum, and iron—are actually a single giant molecule made up of billions upon billions of atoms.

Atoms are so small they don't really "look" like anything—light doesn't interact with them in the same way it does with large-scale structures. But you can think of an atom as a tiny ball of protons and neutrons surrounded by a fuzzy cloud of electrons.

These super-tiny electrons carry a negative electric charge, while the nucleus has a positive charge. It's this electrical attraction that keeps electrons buzzing around their parent atoms. However, when two atoms come close together, the electron is also attracted to the other atom's positively charged nucleus.

If the right combination of atoms comes together, electrons can move between or be shared between the two. This forms what's called a "chemical bond"—and it's the basis for all chemical reactions.

After a bond has formed, the two atoms are stuck together into a molecule. There are two main types of molecule: one made of the same kind of element, and one made of two or more different elements. This second kind of molecule is called a "compound" by chemists.

Some of the simplest molecules are the gases in our atmosphere. Nitrogen and oxygen float around in molecules made up of just two nitrogen and two oxygen atoms. Some of the most complex molecules are the ones found in living things. These so-called organic compounds can be made up of millions or even billions of individual atoms of four or five different elements—usually carbon, oxygen, nitrogen, hydrogen, phosphorus, and sulfur.

A strand of human DNA, for instance, has more than 200 billion individual atoms—and we still need an electron microscope to see it!

So individual molecules are still too small to be useful in making up large chunks of matter like rocks or trees or kitchen cabinets. Fortunately, molecules of the same type often stick together with weaker chemical bonds. The principle is still the same—electric charge attracts the atoms—but because the bonds are weaker, the substance can change its appearance or be broken up quite easily.

Water is a perfect example. An ice cube is made up of billions of water molecules all weakly bonded together. Add a little heat and those bonds break down, and the ice melts into liquid water. Add more heat and the individual water molecules start shooting around at random as steam.

But even at this point you haven't destroyed an actual water molecule. If you want to do that, and crack it into hydrogen and oxygen, you need way more energy and specialized equipment. In fact, this is how we make hydrogen for fuel cells—by cracking water molecules.

From molecules and their chemical bonds comes every physical thing in the world you can touch and use.

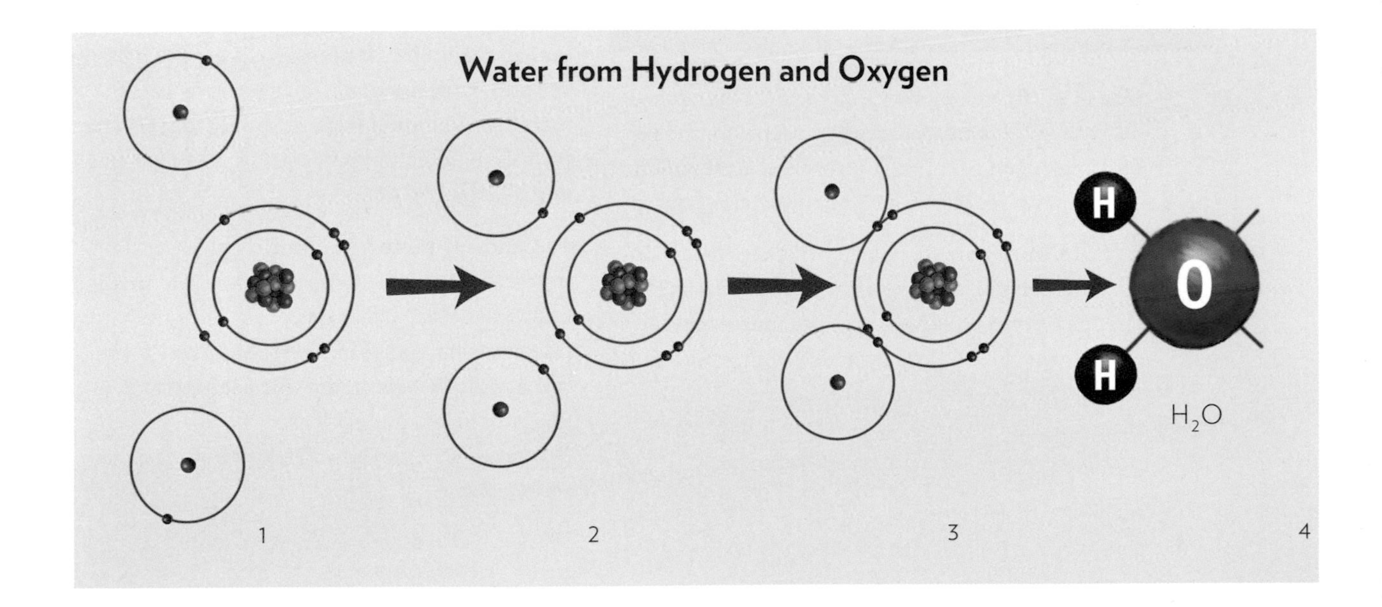

Q:

What exactly is a flame?

Matter can be solid, liquid, or gas. But which one of these applies to flame?

A:

A flame is light emitted from a whole bunch of chemical reactions that occur as a substance burns in a fire. And flames aren't as simple to explain as you might think ....

One of the first chemistry experiments ever done by humans was when some long-ago ancestor took a burning ember from, say, a forest fire, and held it to some dry wood. The wood burst into flame, and so began our long history with fire.

Wait—chemistry experiment? Yes! A fire is a chemical reaction, where heat and fuel combine with oxygen in the air to form new compounds and release heat. It's the heat that humans are most interested in, but fire creates lots of other by-products as well, depending on the fuel used.

Flames are a handy visual cue for us that something is burning. But the chemistry of flame is actually incredibly complex when you look at it down at the level of individual molecules and atoms.

In a small flame, like from a candle, heat makes the fuel—in this case, wax—vaporize. This lets the wax interact with oxygen in the air in a reaction that releases even more heat. We only need to supply some starting heat (a match) to kick off a self-sustaining reaction that lasts as long as there's wax and oxygen to react with each other.

Candle wax is a mix of hydrocarbon molecules that, as it burns, breaks down into smaller molecules. Each break of a chemical bond releases heat. As the chain reaction proceeds, some parts of it get so hot that the electrons in the individual atoms release photons—light particles. These photons let us see the flame.

So really a flame is a glowing zone in a fire made up of millions of chemical reactions. This zone gets pushed and pulled around by the air, making those familiar moving flame shapes we know so well. On the edges of the flame, the reactions are cooling off and slowing down, so the light is less bright and less energetic.

The overall color of the flame is determined by the fuel being burned. Candles burn mostly yellow, but copper burns mostly green. Natural gas stoves burn blue, and pure hydrogen actually burns with an ultraviolet flame that humans can't see. It's all down to the molecules that are involved in the reactions in the heart of the fire.

You'll also notice that flames don't actually touch the thing that's burning. In a wood fire, there's always a tiny gap between the surface of the wood and the flame itself. That's because the flame comes from chemical reactions in the gaseous part of the fire—the wood is supplying a stream of combustible gas via a process called pyrolysis. This causes the wood to char and ultimately break down into ash. All of this is powered by heat.

So really, a flame isn't a "thing" in the standard sense. It's a visible part of a chemical reaction. Without the reaction, there is no flame. You can't capture a flame or put it in a container—though you can contain the reaction that makes the flame.

Chemistry of a Burning Candle

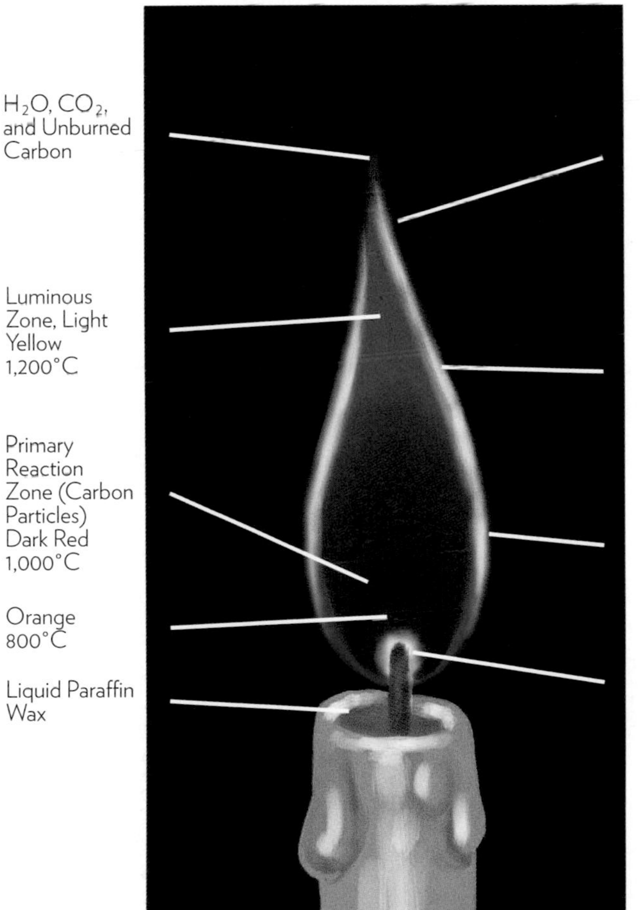

Q:

Why do gasoline engines pollute, while hydrogen fuel cells don't?

The really good thing about running a car on hydrogen is that it is "zero emission"—it doesn't give off any poisonous exhaust. But how is this possible, when regular gas engines are so polluting?

A:

Hydrogen engines get power from a chemical reaction caused by combining atoms together, while gas engines get power from splitting molecules apart. But fuel cells might not be as green as you think ....

The internal combustion engine has done an incredible amount of good for human civilization. These powerful engines haven't just driven us around, they've been used to build cities, generate electricity, and much more.

But they have a downside. They're powered by chemical reactions that produce, along with heat and energy, a whole bunch of nasty new chemicals that can be poisonous or, in the case of carbon dioxide, disrupt the atmosphere.

Gas engines rely on pressure from an explosion to physically move a piston that in turn spins a shaft. In other words, the engine in your gas-powered car uses chemistry to create *force,* which makes the car move.

Enter hydrogen fuel cells—these engines run on a different kind of reaction that instead of directly creating force, creates a flow of electrons, or electricity.

The key difference is that in a hydrogen fuel cell, molecules aren't split apart. Instead, hydrogen and oxygen are combined to create electricity and water. So technically, a hydrogen fuel cell engine does still have emissions—but all it emits is water.

This might not be so harmless, though. When you drive a gas-powered car, the engine uses up the fuel in the tank, making the car lighter. A hydrogen fuel cell makes water by sucking in oxygen from the air—and because oxygen is a bigger atom than hydrogen, the engine will make water equal to nine times the weight of hydrogen it started off with.

Sure, we can just shoot this out a tailpipe as steam, and that's fine if there are only a few hydrogen fuel cell cars on the road. But what if every one of the millions of cars out there was making nine gallons of water for every gallon of hydrogen?

In cold conditions this could even be dangerous, as the water would freeze on the road surface, making it slippery with ice. And water vapor is a much more potent greenhouse gas than carbon dioxide.

One solution could be to trap the water in a tank, which would be emptied when the car refuels with hydrogen. But this means the car would get heavier the longer you drove it—not exactly good news for efficiency or handling! This water would also be very hot—imagine scalding water spraying everywhere in an accident.

What's more, unlike oil, hydrogen doesn't occur naturally (at least, not near the surface of the planet). We need to make it, most commonly by splitting water molecules apart. This takes electricity. If that electricity is supplied via a coal-fired power station, well then the whole "zero emissions" thing goes out the window. The emissions are just happening at the power station instead of from your tailpipe.

All that said, hydrogen fuel cells, even made using today's technology, generate 55 percent less carbon dioxide than an equivalent gasoline engine when you take everything into account. And if the hydrogen is made using solar or nuclear energy, total emissions drop even lower.

What to do with all the water is a challenge—but not an insurmountable one. After all, water is pretty useful!

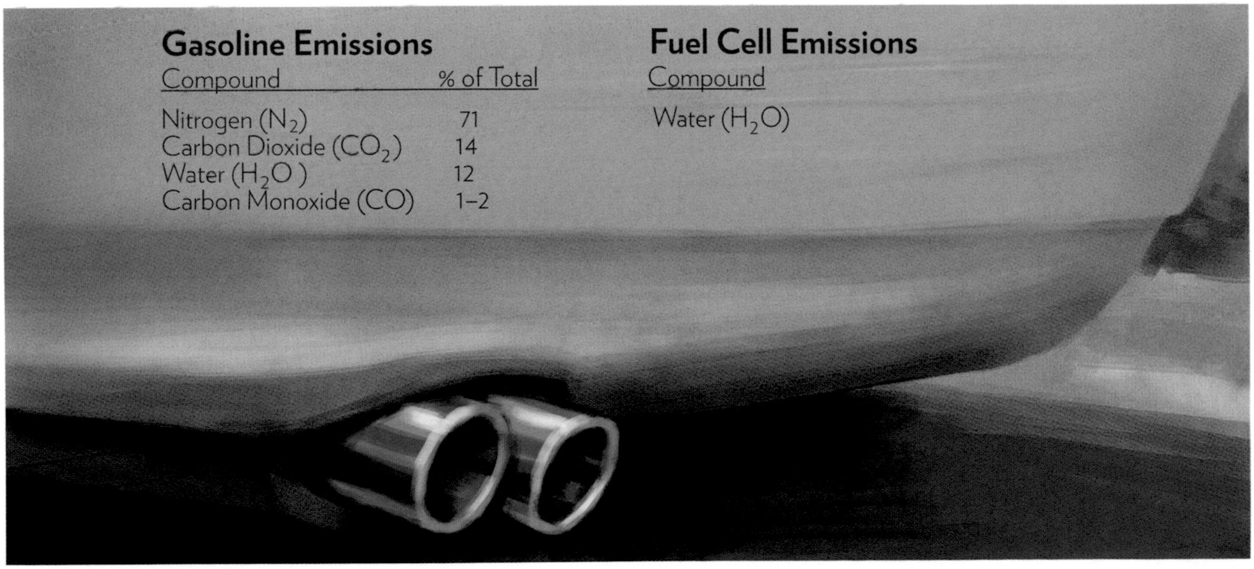

Gasoline Emissions

Compound	% of Total
Nitrogen (N_2)	71
Carbon Dioxide (CO_2)	14
Water (H_2O)	12
Carbon Monoxide (CO)	1–2

Fuel Cell Emissions

Compound
Water (H_2O)

Q:

What's the advantage of cooking our food?

A perfectly grilled steak is much tastier than a raw one, but are there other advantages to cooking food that go beyond flavor and texture?

A:

Cooking causes chemical changes in our food that can make it easier to digest, improve the nutritional content, and even kill off nasty bugs.

Cooking is a unique human ability that, at its most basic, is a set of chemical changes that occur in a substance when it gets heated. We can also "cook" food with acids and by the metabolic processes of some bacteria—we call this fermentation.

The human digestive system doesn't have some of the specialized features we see in other animals. We don't have multiple stomach chambers to break down cellulose in plants, but neither do we have a super-short system suitable only for digesting meat like a carnivore.

Sometime around 700,000 years ago, humans figured out that if they heated food in fire, it became tastier and easier to digest. Cooking massively expanded our diet.

Let's take meat as the first example. While ancient hunters liked to smash open bones and eat the marrow inside, that left an awful lot of the carcass to waste. But the problem with eating muscle—which is what a good steak is made of—is that the fibers of the muscle are surrounded by a material called collagen. Raw, this is hard to chew and not very tasty.

If you're brave, you can try an experiment: cut a tiny piece of raw steak and start chewing. It hardly melts in your mouth—rather, you're left with a sort of nasty, gummy, whitish mess. Which you should then spit out.

Cooking the meat—heating it just enough so the carbon in it doesn't burn and char—breaks down the chemical bonds in the collagen. It also turns the solid fat in the meat into a liquid oil.

Cook too much and the meat gets hard and tough again, but just the right amount and it really does "melt in your mouth." You can chew it and digest it easily.

When it comes to plants, the reason for cooking is slightly different. Plant cells are different from animals because they have a solid cell wall. This wall is tough and hard to break down in our stomach. We *can* eat raw plants and get enough nutrients from them to survive, but if we *cook* the plants the cell walls break down, releasing more nutrients.

In other words, if you cook your food you can eat much less of it to get the same nutrition and energy. And all this is due to chemical changes in the structure of what you're eating.

Apart from giving humans access to a much broader and more dependable range of foodstuff, cooking allowed us to take in more energy in a single meal. More energy allowed our brains to grow bigger and our species to get smarter.

Today, cooking can be a problem because we're so good at packing calories into a single meal we can actually make ourselves obese. Diets that emphasize eating raw food work because they limit the number of calories you can take in.

Badly cooked food, or food made of poor-quality ingredients, can expose you to some nasty chemicals—some of which even cause cancer. But eating raw food has another risk: bugs, specifically bacteria like E. coli, which can kill.

These so-called pathogens can't survive at high temperatures, so well-cooked food is safe food.

Q:

Why do some chemicals explode when you mix them together?

Most explosives need a detonator or an ignition system to set them off, but some chemicals explode when they get mixed. How does this work, and why is it useful?

A:

Chemicals that explode on contact are called "hypergolic," and they contain oxygen so they don't need air or heat to react. They're most useful as rocket fuel ....

Explosives like TNT and dynamite are plenty dangerous, but at least you need to add extra energy before they'll go off. Even very unstable explosives like nitroglycerin are safe if you keep them still, dry, and out of sunlight.

Some chemicals, though, are so reactive that they'll explode if you even mix them. This kind of reaction is called a "hypergolic reaction," and it's ideal for use as rocket fuel.

Explosives work because the chemicals that they're made up of are packed with lots of chemical bonds. These bonds store energy, and when the bonds break all that energy is released.

Most explosives need a little encouragement, usually in the form of fire via a fuse. More modern explosives use an electric charge to provide a jolt of energy. And very unstable explosives like nitroglycerin will go off when just a tiny amount of extra energy is added—like dropping it onto a hard surface.

Hypergolic explosives are different because they're made up of two ingredients. One chemical is the fuel, and the other is the oxidizer. All explosives need oxygen to go off, but standard explosives get that oxygen from the air.

Hypergolics bring their own oxygen to the party, and because of that they don't need an ignition system. You just mix them together and BOOM!

Where this is most useful is in rocket engines. Up in space, the fewer parts you have in an engine, the more reliable it will be. A hypergolic engine just needs a couple pumps and a chamber where the explosives—called propellants—can mix.

To stop the engine, just turn off the pumps. The chamber uses up all the propellant and the engine stops. Easy!

One of the really big advantages of hypergolic explosives is that the individual ingredients can be stored as liquid at normal temperatures. Compare this to a rocket engine that uses liquid oxygen—this type needs a so-called "cryogenic" system to keep the oxygen very cold and liquid.

So hypergolics are simple and reliable, but they're less powerful than other explosives. Unfortunately, one or other of the ingredients also tends to be very acidic or even carcinogenic, so storing and transporting them safely is tricky.

In your day-to-day life you probably won't encounter hypergolic chemicals, like Aerozine 50 or nitrogen tetroxide. But you might encounter other chemicals that react when they contact each other, such as certain types of glue that come in two tubes. When you mix the tubes, the glue sets.

Improper storage of chemicals can sometimes lead to hypergolic explosions, as anyone who works in a chemical lab will tell you. Certain chemicals, especially those with lots of oxygen in them, will be kept in separate cupboards. They might not *explode* when they contact the wrong stuff, but they could get very hot and start a fire.

Another place you can see this kind of reaction is in a glow-stick. When you get one out of the box, two chemicals inside are kept separate because one is in a thinner tube inside the other. When you "crack" the glow-stick, the inner tube shatters and the chemicals mix. The reaction is only strong enough to produce light, and the chemicals aren't oxidizing like in an explosion, but it's basically the same idea.

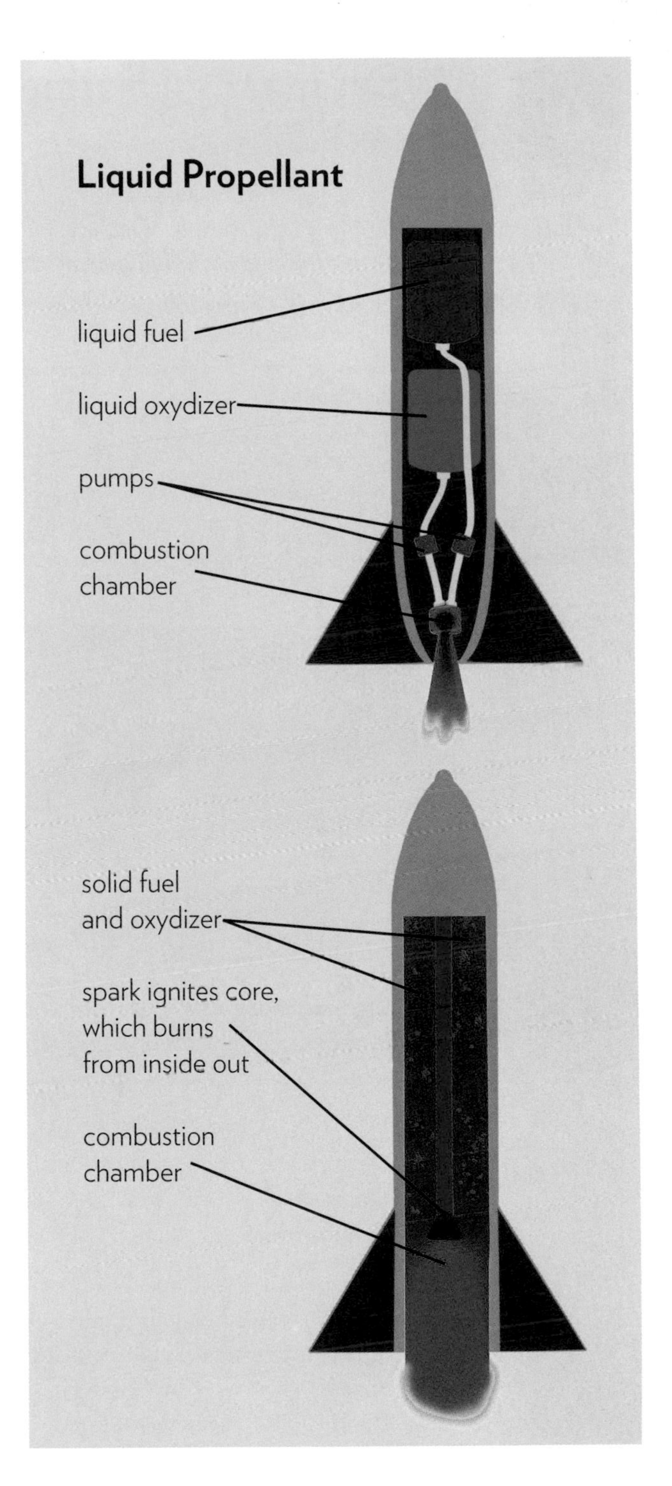

Q:

What makes gasoline such a good fuel?

When it comes to weight, ease of transport, and power, gasoline seems to have other fuels beaten hands-down. Why is it so good?

A:

Apart from containing lots of energy, gasoline is also very stable and easy to store. But when you take cost out of the equation, gasoline might not be so great ....

The main reason that gasoline is the "best" fuel you can use in your car right now is ultimately the fact it's available everywhere and doesn't cost very much.

However, there is some scientific basis to the claim that gasoline is a very good fuel for internal combustion engines.

Most vehicles and airplanes today still work on the principle of exploding some kind of oil-derived fuel to create hot air that expands quickly. This expanding air forces a piston to move or makes a turbine spin. This movement—or kinetic energy—is then translated into forward motion. In a car, it goes into a transmission system that turns the drive wheel. In a propeller plane, the spinning propeller pulls the aircraft through the air. And in a jet plane, compressed air shoots out the back of the engine, pushing the plane forward.

There are many other chemicals that create more energy from exploding than gasoline, but you have to transport them at low temperature or in pressurized containers. Gasoline is great because you just pour it into a tank. In fact, as a liquid gasoline is relatively hard to ignite.

When gasoline evaporates into a vapor, *then* it becomes very explosive. This is how most engines work—they squirt the gasoline into an ignition chamber as a mist and add a spark. Boom, the air in the piston gets super hot, the piston is forced down, and energy is transferred.

This system isn't very efficient, though. While about 20 percent of the energy is converted into movement, the rest is wasted as heat and sound. But for the whole of the twentieth century that didn't really matter—the amount of useable energy from gasoline was so huge, we didn't really care that most of the total energy from the burning of the gas gets wasted.

A modern electric engine with top-shelf battery technology, on the other hand, can be up to 90 percent efficient. That is, of the energy being put out by the battery, 90 percent of it goes into the motor to turn the wheels.

Again, though, the sheer amount of power a gallon of gasoline can put out means its lack of efficiency almost doesn't matter. The key here is energy density—there is a *lot* of energy in a gallon of gas.

Gasoline is refined from crude oil, and it's a super-complex hydrocarbon packed with lots of chemical bonds. Breaking these bonds releases energy. Compared to other fuels, a well-tuned gas engine actually burns really clean—it produces just carbon dioxide and water.

Unfortunately, we've gone and made one billion cars. With so many, we're using up all the oil and pumping out way too much CO_2.

The thing is, gasoline is only the best because it's been easy to make over the last 100 years. And it's only easy to make because there's a global industry making it. It took gasoline a long time to really catch on as a fuel—about 50 years before it was easy to get anywhere.

The significant downsides to gasoline are now starting to be felt. It's toxic, it's hard to clean up if it spills, and that inefficiency is really starting to bite. Time for electricity to have its day!

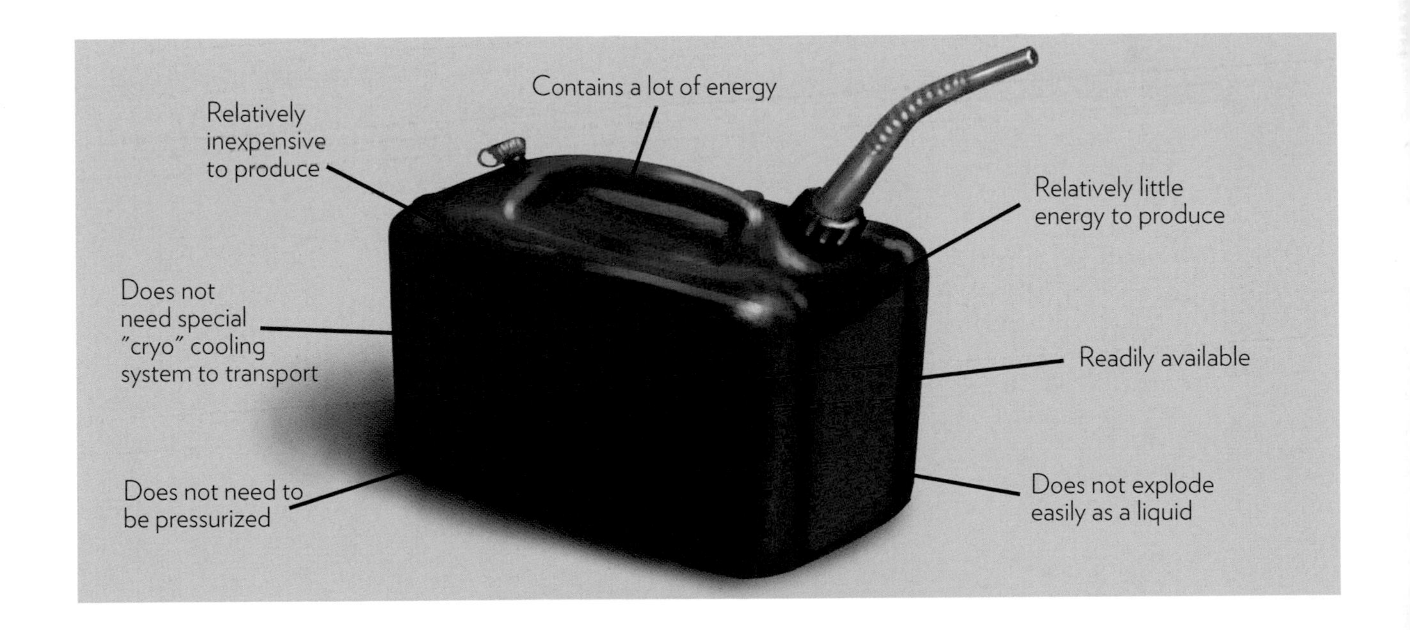

Q:

Why is smell our weakest sense?

Compared to sight, hearing, touch, and even taste, smell is much more subtle. What makes smell our least useful sense?

A:

Smell relies on sensors picking up specific chemicals that have to waft through the air in just the right direction. But we don't yet fully understand how the sense of smell works ....

Scientists call smell the "olfactory" sense and, like taste, it relies on a sensor cell in the body coming into direct contact with a chemical. The sensor cell, called a "chemoreceptor," detects the chemical and fires a nerve, which sends a signal to the brain. Congratulations—you just smelled a rose!

The chemicals that we smell are of a particular type called "odorants." You're not sniffing up a fleck of the rose's petal, for instance, but rather a chemical the rose deliberately emits to attract bees and birds.

Other odorants float off objects just because they're exposed to the air. The general smelliness of an object depends on how many odorants it puts off. Rotten meat puts off a lot, but a piece of glass puts off very few.

Heat affects how many odorants drift off an object. Concrete is normally not very smelly when it's cool, but has a distinct odor when heated up by the Sun. As objects get hotter, the chemical bonds holding molecules to their surface break, allowing the chemical to float off into the air.

Other substances react constantly with air, creating new molecules (rusting metal is one example). Some building materials exhibit so-called "out-gassing" in which chemicals from the manufacturing process escape. New carpet smell, new paint smell, and even new car smell are examples of this. The smells fade as the chemical reactions slow down over time.

The exact way a bunch of chemicals hitting your nose receptors gets turned into a sensation of smell isn't yet totally understood. It's complicated, because your brain doesn't just react to a single chemical at a time. When you smell a roaring fire, small particles of carbon will hit your receptors, but other chemicals from the wood and the sap will also set off different receptors. Your brain actually receives multiple activation signals at once—one for the carbon atom, dozens more for the complex organic compounds from the wood. And if your fireplace or burner has a metal chimney or metal grate, tiny particles of metal will also add to the overall smell of the fire.

One of the more widely accepted theories suggests the brain has a "chemotopic map." It's as if our brains come pre-programmed to identify certain combinations of chemicals as certain smells. Your sensors pick up all the different molecules, and then nerves carry that information to the brain, which matches the pattern against the chemotopic map.

We actually have separate smell processors for each nostril, so it's possible to smell two things at once by, say, putting a strawberry-scented chapstick under one nostril and a peppermint under the other.

The reason smell is so weak compared to other senses is simply because it needs chemicals to drift by. Our bodies are constantly bombarded by photons from light sources, so sight works easily. Touch works by direct contact with what we're sensing. And taste is similar to smell in that it detects chemicals, but since we're putting stuff in our mouth it has a lot more chemicals to react with.

When you take a breath, only a tiny proportion of the air is made up of odorant chemicals. Of course, if you breathe air from a confined space such as, uh, a recently used bathroom, the density of odorants is much higher, and you'll probably start wishing your sense of smell was even weaker ....

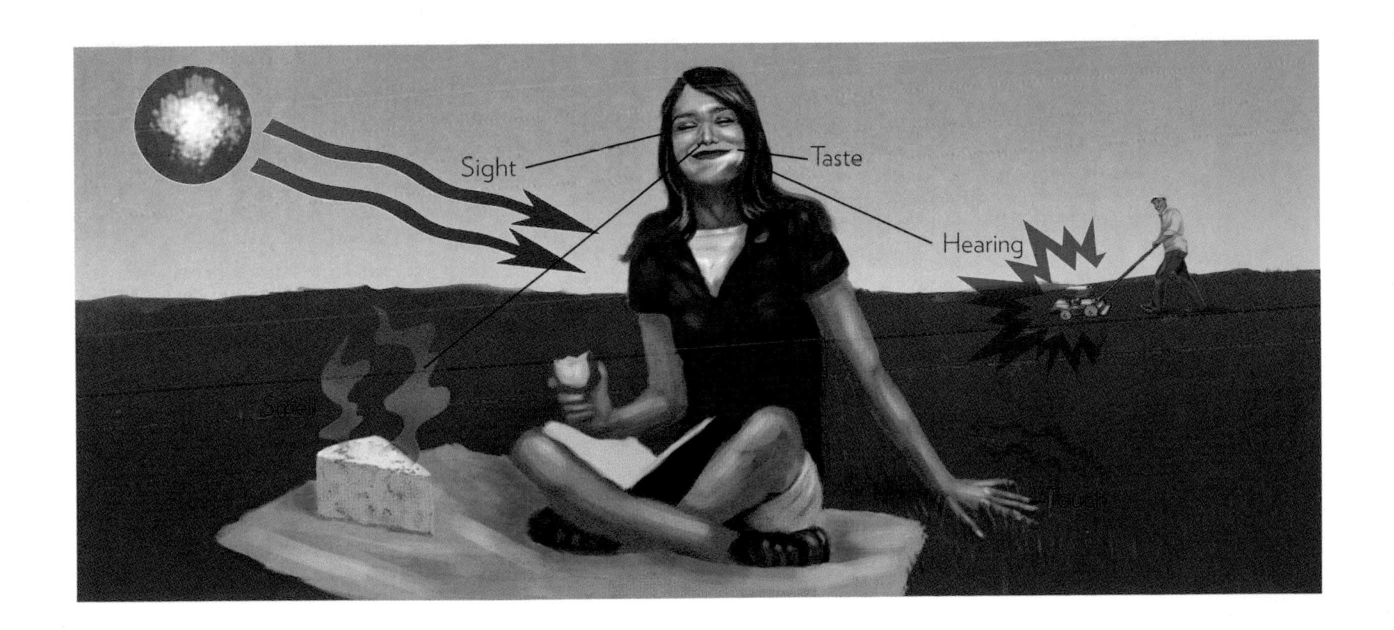

Q:

How does light "charge up" glow-in-the-dark stickers and toys?

Glow-in-the-dark is a big hit with kids, but how does this substance trap and store light?

A:

Glow-in-the-dark products are painted or mixed with a chemical called a "phosphor." This substance absorbs energy and then releases it slowly as a dim green light.

The easiest way to turn a toy from cool to awesome in the mind of a young child is to make it glow in the dark. Typically white with a faint green tinge under normal light, phosphors will glow a dim ghostly green once you switch the lights off. Over the next 10 to 20 minutes, it slowly fades into darkness.

First things first: you might have heard that this stuff is radioactive. It isn't ... but it used to be mixed with radioactive material, traditionally radon. That's because the radon would supply a continuous stream of energy via radioactive particles that would keep the glow-in-the-dark paint glowing.

The glowing toys and stickers you can buy today are made with either zinc sulfide or strontium aluminate. These aren't the kind of chemicals you'd want to eat, but they're pretty stable and not very toxic in normal use.

When you charge phosphors with energy, they release that energy slowly in the form of visible light. So if you keep your glow-in-the-dark toys in a closet, they won't glow. Hold them up to a light for just a few moments, though, and they'll collect enough energy for 10 minutes or so of bedtime fun.

There are lots of different chemicals that act as phosphors, but zinc sulfide is commonly used because it's cheap and you can charge it up using normal light. In fact, pure zinc sulfide doesn't glow—chemists add a tiny amount of copper to get that familiar greenish color. If silver is added, it glows bright blue.

The exact mechanism by which phosphors work is very complex, but essentially adding extra energy slightly destabilizes the chemical. To fix this and return to a stable state, the chemical emits a photon—which we see as light.

The wavelength or color of the light depends on the types of atoms mixed in with the chemical. Once enough energy has been thrown out of the chemical in the form of photons, it stops emitting light and the glow fades.

Zinc sulfide doesn't just glow after absorbing normal light. It's also used in medicine because it glows after absorbing x-rays.

It's true that once upon a time watches had radioactive radon or even uranium paint. These radioactive materials emit tiny particles, which the zinc sulfide absorbs. This makes the watch glow constantly, even after hours of darkness. Because the radioactive paint was encased in a stainless-steel watch and behind glass, the chances of the watch irradiating your arm were very low.

Still, public opinion is a powerful thing, and today radioactive paint is harder to come by. But you can find tritium (a radioactive form of hydrogen) in some illuminated gun sights and also in the instruments of some airplanes.

The more modern and expensive chemical, strontium aluminate, gets used in fancier glow-in-the-dark stuff today. It works in the same way as zinc sulfide, but it can store more energy and glow longer—about 10 times as long. And it glows 10 times as bright!

An Atom of a Phosphor Molecule

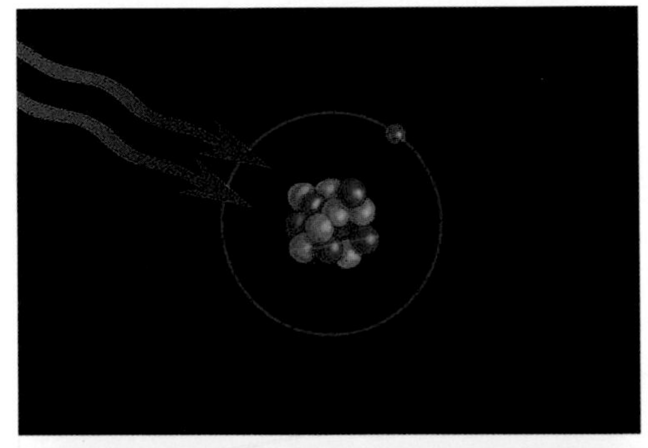

Incoming Light Energy

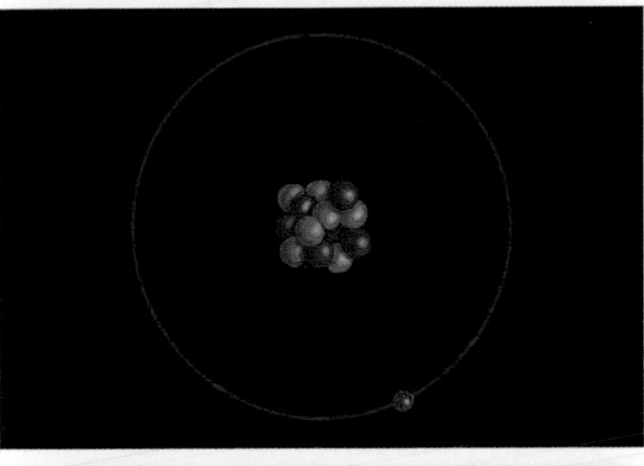

Energy Stored

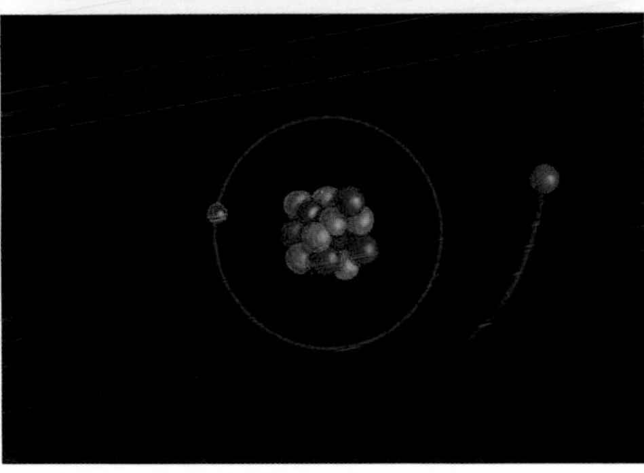

Energy Released

Q:

How does our sense of taste work?

Without our ability to taste food, the world's fine dining industry would be in big trouble. But how does taste work and why are some foods "tastier" than others?

A:

Sensors on our tongue detect certain chemicals, but taste is more complex than that. Our sense of smell, sight, and touch all work together to give us the full food experience.

Explaining the mechanics of taste is pretty straightforward: the human tongue is covered with tiny buds. Each taste bud has a pore in the top, which picks up chemicals in our food. The chemical stimulates a receptor cell that sends a signal into a nerve. The nerve contacts the brain, and we register the taste.

Different sensor cells cluster at different spots on the tongue, giving us five basic tastes. Four are obvious—sweet, sour, salty, and bitter. The fifth was only properly described in 1908 and uses the Japanese word *umami*. It picks up that hard-to-define sense of deliciousness you get from some savory foods like cheese, some meats, and soy sauce.

When you chew food, chemicals are spread all over your tongue and are picked up by multiple taste buds. So some food can be sweet and sour at the same time, or salty and delicious, or some other combination.

But our tongues also have other sensors in them that aren't, strictly speaking, taste buds. Instead, our sense of touch can communicate a lot about a food's flavor.

Alcohol, chili, and other spicy foods stimulate the same nerve pathways in our tongue and mouth that would activate if we really were licking something that was actually hot. Chili more or less tricks your mouth into thinking it's on fire.

We get the opposite effect from minty foods, which stimulate the cold-detecting nerves.

And if you ever get a metallic taste in your mouth, that's your nerves picking up a very faint electric current or flow of electrons from, say, metal fillings or even iron-rich blood.

To get the full sensory experience of eating a favorite food, we also need our sense of smell and sight. All four senses—taste, touch, sight, and smell—send nerve signals that combine in our brain into a complete sensation.

Experiments have been done in which people are blindfolded or have their sense of smell disabled, and it has a definite effect on the perceived taste of food. Without the brilliant redness of a tomato, the flavor seems less rounded or complete.

Having a sense of touch in your mouth is extremely important, because without it you wouldn't be able to tell if a food was hot or cold, and you'd potentially damage your mouth from burning or even freezing. As a happy side effect, temperature sensors combine with taste sensors to enhance the flavor of various foods. For example, coffee tastes better either quite hot or quite cold; at room temperature, it's not so good.

Why is temperature important? Because some foods have a complex chemical mix that, while it includes yummy tastes, can also include less pleasant tastes—especially bitterness.

If the food is hot, the nerve signals from your heat-detecting cells can overwhelm or drown out the signals from the bitterness-detecting cells. The same goes for very cold food.

Salt is another good way to drown out bitterness, and people who are sensitive to bitter flavors will often put a seemingly crazy amount of salt on their food.

Because taste is so complex and subjective, humans can be incredibly varied in their, well, tastes. Some people *do* like room-temperature coffee. Others like hot soda. And then there are the pickles on fast-food burgers ....

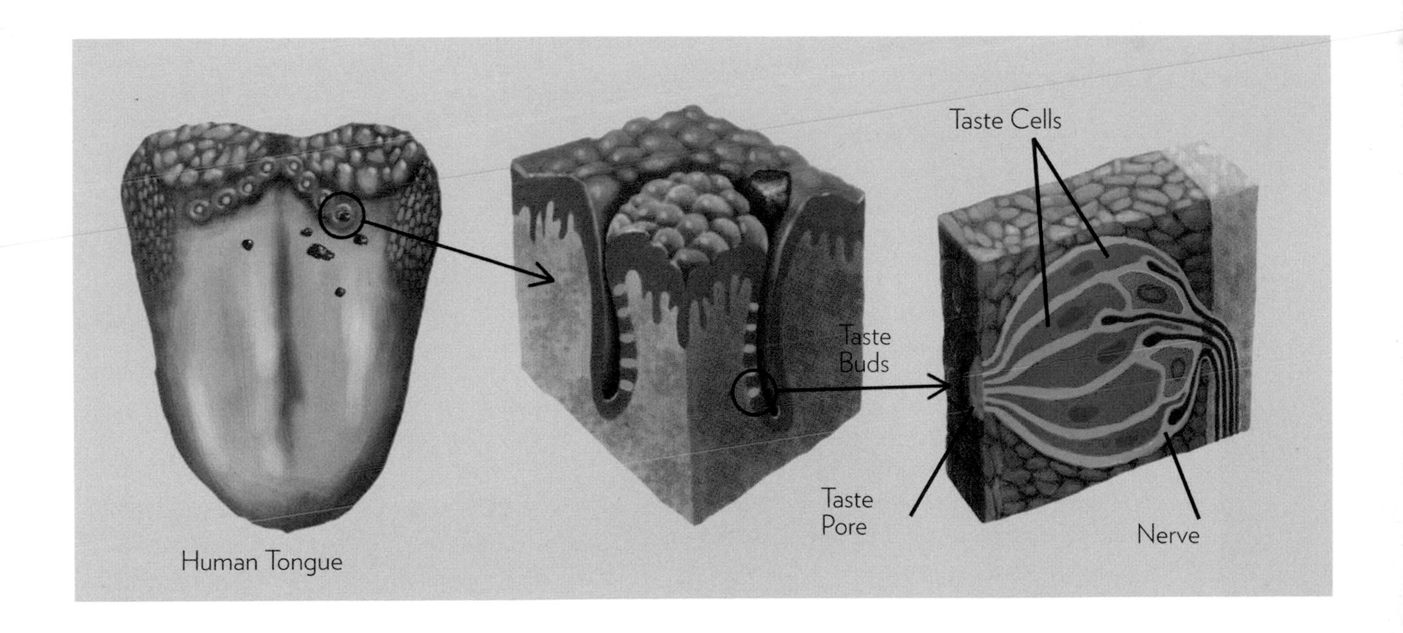

Q:

Would it be possible to freeze the air solid?

If the Sun was extinguished or Earth drifted off into space, would the atmosphere eventually freeze into a giant block of ice?

A:

All the elements in our atmosphere can freeze, but they freeze at different temperatures. In fact, it's for this reason that there's life on Earth at all ....

The "scientific" answer to this question presents one of those times when we have to argue about definitions. Yes, every element in the air—mostly nitrogen, oxygen, carbon dioxide, and a bunch of other stuff—can freeze at a low enough temperature. But these elements all freeze at different temperatures. As we lower the temperature of the atmosphere, they'll turn into different kinds of snow and fall to the ground.

So the question is: are we really freezing the air, or just the individual components of the air, one by one? It's hard to think of a situation where the atmosphere would freeze so quickly that any sample you took of the resulting snow or ice would have the same mix of chemicals in it as a sample of the atmosphere does now.

Compared to the planet itself, our atmosphere is quite thin. If you really could somehow freeze it instantly, it would condense into a layer of snow or ice only about 350 feet (107m) thick! This may have in fact happened on Mars—the Red Planet's polar ice caps, which are made of carbon dioxide, could be all that remains of its ancient atmosphere.

Earth's atmosphere isn't very vulnerable to freezing, though. That's because we have a mostly nitrogen atmosphere, and nitrogen has an incredibly low freezing point.

We can condense nitrogen into a liquid using industrial processes, but we have to store it at -321°F. If you want nitrogen to freeze into a solid, you need to drop the temperature to -346°F. Oxygen is even harder to freeze. It condenses into a liquid at -297°F and freezes at -368°F.

The next biggest component of our atmosphere is argon, but it makes up less than 1 percent of the total. It freezes at about -308°F.

So for our atmosphere to freeze, the Earth would need to get incredibly cold. Turning off the Sun would do it—our surface temperature would eventually stabilize at about -400°F thanks to heat leaking out from the core. Cold enough for nitrogen snow, even.

There is one chemical in our atmosphere that freezes out all the time, though—water. Gaseous water is an essential part of the air, and a vital part of the so-called water cycle.

When water drifts high into the atmosphere, it cools and condenses into tiny liquid water droplets. Most often, these stick together and form clouds. When a cloud gets enough water in it for the droplets to be too heavy to float, they fall out as rain.

Snow clouds are just normal water clouds that are much colder, so the water freezes into ice. Again, these solid crystals grow larger as they combine, until they get heavy enough to fall as snow. And there's an intermediate form where liquid water hits cold air and freezes very rapidly—that gives us hail.

The fact that Earth is just the right temperature for water to exist as a liquid, a solid, and a gas is one of the keys to life. Without an atmosphere cool enough to condense and freeze water, all our water would slowly escape into space. This might have happened to Mars.

If the Sun Were to Disappear or Go Out All at Once

Oxygen turns into liquid -297°F

Argon turns to liquid -302°F

Argon freezes into a solid -308 F

Nitrogen turns to liquid -321°F

Nitrogen freezes solid -346°F

Oxygen freezes solid -368°F

Earth's eventual surface temperature -400°F

Q:

How does oxygen actually give me energy to survive?

Humans can only live for a few minutes without a constant supply of oxygen. Why is it so important, and how exactly does this volatile gas give us energy?

A:

Oxygen is a reactive chemical that is essential for cellular processes to create energy. In fact, we use oxygen in a very similar way to a gasoline engine, and our "exhaust" is nearly the same, too ....

With the exception of a few types of bacteria and other simple microbes, all life on Earth uses oxygen. Even plants use it, taking it in through their roots.

Oxygen is a very useful chemical for driving reactions, because oxygen atoms readily accept electrons from other atoms. Getting energy out of a bunch of chemicals is all about breaking so-called high-energy bonds. The way these bonds actually work is pretty complicated, but basically reactions can break bonds by pulling electrons around.

Oxygen is good at pulling electrons. We often talk about "oxidization" as a chemical process, and it can refer to everything from burning wood in a fire to igniting gasoline or even rusting an iron bar. All these things are chemical reactions in which oxygen is part of a process of breaking chemical bonds and releasing energy.

Humans are a "eukaryotic" life form. That means we're made up of trillions of cells, and each cell has a nucleus (with a few exceptions that have none, such as our red blood cells). Our cells have a whole bunch of structures inside them called "organelles" that do different jobs, including the absolutely vital task of providing energy for cellular processes.

Cells need to do a few different things. They need to grow and divide, of course, but some—such as muscle cells—need to generate physical movement. Other cells, like nerve cells, need to generate electrical currents. All these things require energy.

Our circulatory system provides our cells with nutrients from our food, and our blood supplies cells with oxygen from the air.

Cells use oxygen molecules in a combustion reaction to create a substance called adenosine triphosphate, or ATP. This chemical gets passed around inside the cell and actually carries the energy the cell needs to work. We use oxygen to make ATP, and structures in our cells use the ATP to get energy.

Our supply of ATP gets recycled throughout the day. At any given time, the average person has just 8.8 ounces (260ml) of ATP in their body, but they'll generate their entire body weight in ATP over the course of the day. That's how fast our body cycles through the stuff.

Strangely enough, when we use oxygen for energy we produce waste that's very similar to the exhaust of a gasoline engine: carbon dioxide and water. We breathe the CO_2 out, and use the water for various biological processes.

So humans are an "internal combustion" life form, just like a car has an internal combustion engine. Of course, our combustion of oxygen is more complex than just applying heat to a hydrocarbon such as gas. Technically, there is a whole string of reactions that convert sugar and other nutrients from our food into energy, but oxygen is a vital part of those reactions.

Some life forms don't use oxygen in their cellular processes, substituting it with a different chemical such as sulfate or nitrate—though it's interesting to note that both of these chemical compounds do contain oxygen. But only so-called aerobic life uses pure oxygen to drive reactions in its cells.

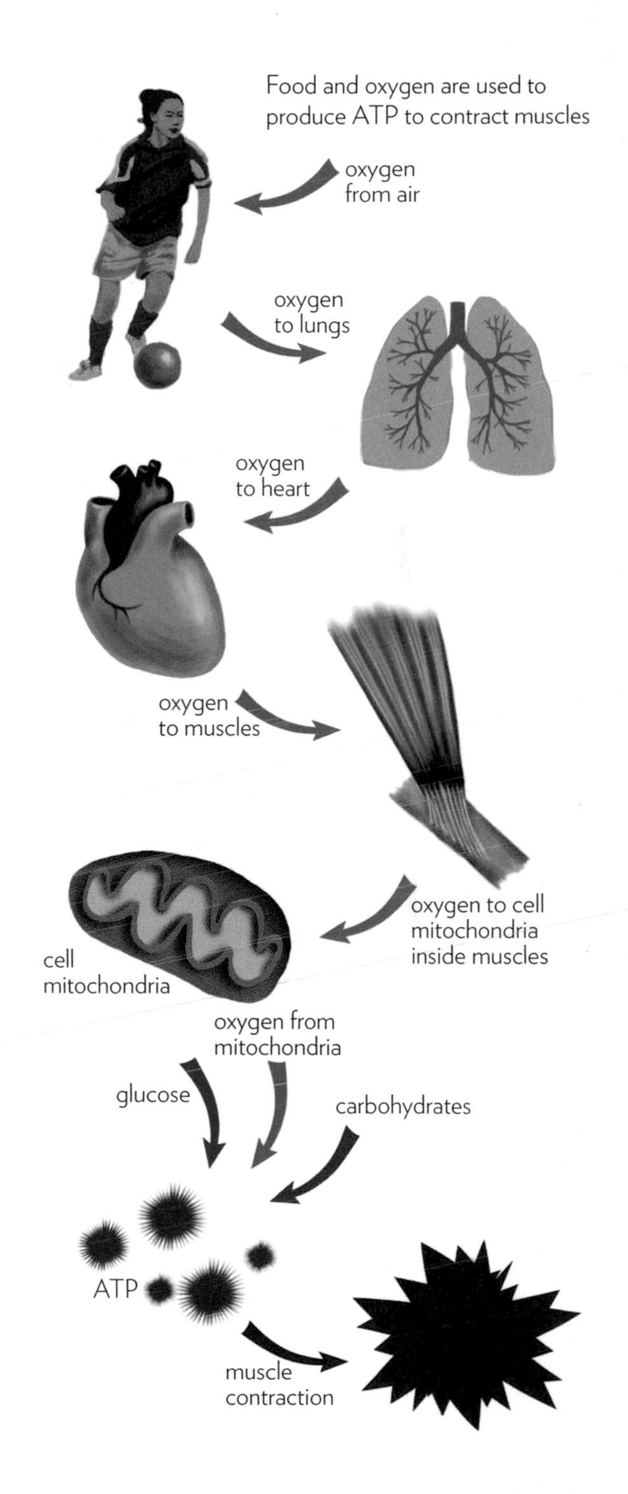

Q:

Why is carbon monoxide in car exhaust so dangerous?

It is a tragedy that people are able to harm themselves by simply breathing car exhaust in an enclosed space. But how does this actually kill us?

A:

Carbon monoxide reacts in our blood in the same way as oxygen, but can't be used by our cells. It literally clogs up our blood, and we suffocate.

Every cell in our body needs oxygen to run chemical reactions that in turn produce energy. Only with a constant supply of energy can we survive.

Our brains in particular need a lot of energy to generate the electrical impulses that form our thoughts and to process the information coming from our senses.

Oxygen is carried to our tissues by special cells in our blood. Human blood looks like a thick red liquid, but the actual liquid part is a pale yellow. The red color comes from a couple trillion tiny, dish-shaped cells called red blood cells—or *erythrocytes* if you want to be scientifically precise!

The average person makes about 2.4 million red blood cells every second, deep inside their bones, in the stuff called marrow. The cells are red because they're full of a protein called hemoglobin. This contains lots of iron and as a result is really good at binding oxygen. Because we pack oxygen into our red blood cells, we can carry 70 times more oxygen than if the gas was just dissolved in the liquid part of our blood.

Unfortunately, hemoglobin doesn't just bind oxygen. It's also really good at binding a chemical called carbon monoxide. The problem with carbon monoxide is that our bodies can't use it to make energy for our cells.

Normally what happens when a cell encounters a red blood cell is that it sucks the oxygen out of the blood. Then the red blood cell circulates back through the body to the lungs, where it picks up more oxygen and the process repeats.

When a red blood cell becomes clogged with carbon monoxide, no other cell can get rid of that chemical, so the red blood cell can no longer transport anything. It just circulates around doing nothing until the carbon monoxide pops out—this can take as long as five hours!

If you breathe a little bit of carbon monoxide, some of your red blood cells get clogged, and you might experience some symptoms such as lightheadedness, confusion, headache, or vertigo. If you're unlucky enough to live somewhere there's a slow carbon monoxide leak, and you breathe a little bit of the stuff every day, you might become depressed and suffer some memory loss. This happened to a number of early Antarctic explorers when they spent whole winters shut up in a tiny hut with a badly ventilated stove.

Curing carbon monoxide poisoning is pretty simple: doctors will stick you in a so-called hyperbaric chamber filled with a much higher than normal percentage of oxygen. Or they might just give you a breathing mask attached to an oxygen tank. By filling your lungs and blood with a higher-than-normal amount, this treatment can saturate your blood with enough oxygen to keep you alive while the carbon monoxide gets flushed out of your system.

Sadly, people who are determined to harm themselves can shut themselves up in a garage with a car engine running and breathe heavily concentrated carbon monoxide from the exhaust.

Older cars made this easy, but a sophisticated, well-tuned modern engine with a catalytic converter produces very little carbon monoxide. But people can still inhale a harmful dose.

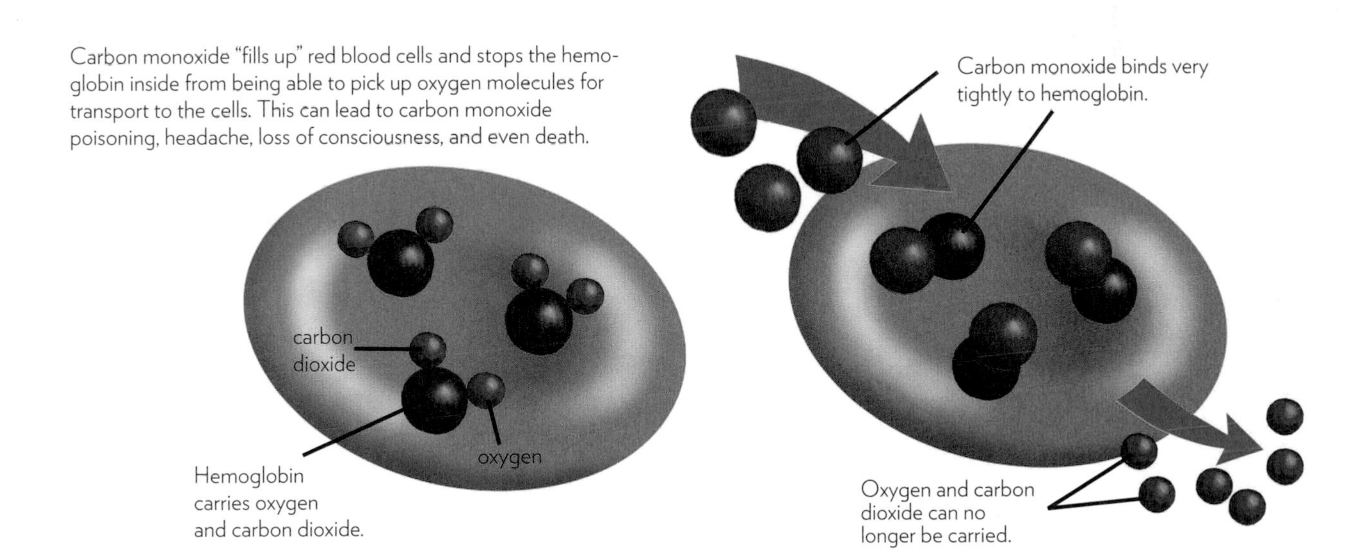

Carbon monoxide "fills up" red blood cells and stops the hemoglobin inside from being able to pick up oxygen molecules for transport to the cells. This can lead to carbon monoxide poisoning, headache, loss of consciousness, and even death.

Q:

How is it possible for food companies to make artificial flavors?

Even though a candy company might claim their latest creation has "no artificial colors or flavors," we know there isn't actual apple or strawberry in those gumdrops. But how do these fake flavors work?

A:

Our sense of taste reacts to certain chemicals, and chemists can make other similar chemicals in the lab—but our tongues are rarely completely fooled ....

Our tongues are covered with thousands of tiny taste buds that have chemical sensors in them. When a certain type of chemical hits the sensor, a signal is sent to our brain via a nerve and we perceive a flavor.

From childhood we learn to recognize certain flavors, mostly through experience. We know what apple tastes like, what mint tastes like, and so on.

The thing is, those tastes come from fairly simple chemicals, and it just so happens there are many possible chemicals that, to our taste buds, look more or less the same as the real flavors.

There's a special kind of chemical compound called an "ester." Esters are the result of mixing an acid with an alcohol and end up in all sorts of crazy stuff from plastic (such as polyester) to explosives like nitroglycerin!

Most organic substances that we like to eat, especially fruits, contain esters that are part of what our tongues detect as flavor. So chemists only have to find a chemical that resembles the real esters in, say, an apple in order to create a fake flavor that tastes a lot like apple.

This is how candy works. Most candy is just sugar embedded in something like gelatin, with an artificial color and flavor added. By adding a few drops of an ester such as "ethyl butyrate," candy makers can make you think the candy tastes sort of like banana, pineapple, or strawberry.

Wait—how can one ester make three different flavors? Because our sense of taste is about more than just detecting a single chemical. If the candy is shaped and colored like a strawberry, your brain will think ethyl butyrate tastes more like strawberry. If the candy is yellow and pineapple shaped, well, you get the idea.

That said, very few people would agree that fake candy flavors taste exactly like the real thing. In fact, most people would say candy only "sort of" tastes like the real thing.

That's because a real strawberry contains many different chemicals, including lots of different types of esters. It has chemicals that carry a bitter flavor along with sweet and chemicals that make other cells in our mouth react to tartness or astringency—depending on how ripe the strawberry is.

A real-life flavor is very complex and made up of lots of chemicals. What's more, the texture and smell of a real strawberry affect your perception of its flavor as well.

There are people whose job it is to mix chemicals to make artificial flavors as convincing as possible. Some of the banana flavors can be quite good because real banana is heavily dominated by a single ester called isoamyl acetate. Flavor-making chemists will even add perfumes and other chemicals you're meant to smell rather than taste, because smell is so important in the perceived flavor of a food.

Some artificial flavors don't even try to mimic real-life flavors. Certain brands of gum and energy drinks have flavors that are described by not much more than their color. People still seem to like them, though ....

Q:

Why does unhealthy food make me fat?

That eating too much leads to obesity seems pretty obvious, but why is it that so-called unhealthy food makes me fat faster and more easily?

A:

We put on weight when we consume more calories than we expend through physical activity. Unhealthy food has more calories but, crucially, it's easier to eat than other food ....

Because an animal's supply of food isn't always assured, evolution has come up with a few mechanisms to deal with times of famine and times of plenty. Humans and many other animals have the ability to store excess chemical energy in the form of specialized cells—fat cells.

Our digestive system and metabolism work together to decide that a certain portion of the calories we eat aren't needed for the day's activity and for growth. So those calories are sent to our tummies or thighs and turned into fat cells.

Fat is like a battery. We charge it up and then use it as needed. Without fat, our metabolism would have a hard time taking in exactly the right number of calories for a day's living. Fat makes eating easier.

There have been fat and even obese people throughout history. For whatever reason (usually extreme wealth and power), these people have had access to way more calories than they need, and this excess energy gets stored as fat. If you eat enough of even very healthy and low-fat food, you can become obese.

When he died at the age of just 55, King Henry VIII was famously obese and plagued with health problems. But because he lived in the late fifteenth century, everything he ate was farmed organically. Forget artificial flavors and pesticides—he ate nothing but the finest natural produce ... cooked excessively and stuffed with heavy cream and pure fat. He also had an ulcerated leg wound, a damaged immune system, and possibly untreated type 2 diabetes, which didn't help. But it does go to show that supposedly healthy food can make you fat if you eat too much of it.

That said, unhealthy food—by which we mean heavily processed foods and foods with lots of additives like pure sugar—can get you to Henry VIII condition much faster. And you don't have to be the richest man in England to afford it, either.

The average human needs about 2,000 calories of energy to get through a normal day without having to use up any stored fat. Individual people might need a little more or a little less, depending not only on how big they are but also on the mix of bacteria in their gut. New research suggests that some people are better than others at turning energy into fat, which means they will get fatter on slightly fewer calories.

Modern processed food is incredibly calorie dense compared to the food our ancestors ate, and on top of that it's very cheap. For just a few dollars you can buy a pizza with 1,500 calories, for example. Add in a soda and some garlic bread and you've eaten a day's worth of calories in a single meal.

Unlike a roasted swan stuffed with 4 pounds of goose liver pâté (something Henry VIII would have loved), a pizza is also very easy to eat. This is a key characteristic of unhealthy food—you can eat it fast, and it doesn't fill you up.

All these things add up to a dangerous whole: cheap food with lots of calories that's easy to eat. The result is obvious. You'll get fat.

Why Fast Food Makes You Fat

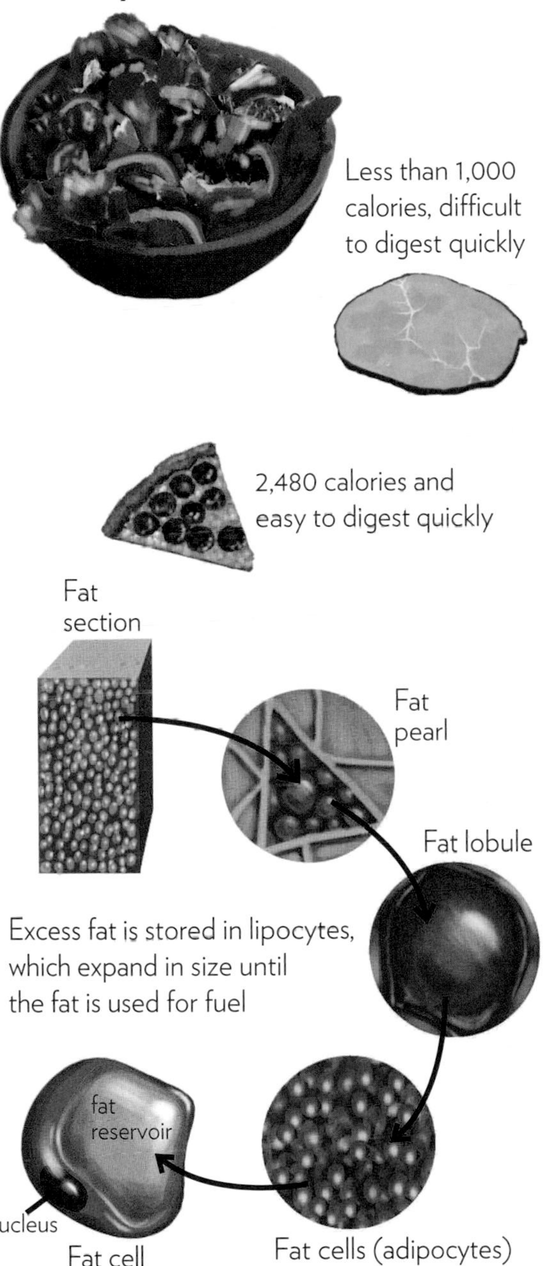

Q:

What makes some things brittle, instead of just hard or soft?

Some substances are very hard, others quite soft and malleable. But then there are those that are brittle: hit them hard enough and they shatter. Why does this happen?

A:

Brittle substances have their atoms bonded in a crystal lattice, and if we bend or stretch that lattice enough, the bonds snap—often quite violently!

For this answer, we're talking specifically about pure substances, things like diamond or copper that are made up mostly of one kind of chemical compound. More complex objects that are made of different substances, like wood, which has cells and other structures, break and shatter according to different rules—mostly they break where the structures inside them are weakest.

A pure substance like diamond has a very regular internal structure. If you zoom right in to the atomic level, you'll see that diamond is made up of trillions of carbon atoms arranged in a particular shape. Each carbon atom is attached to four other carbon atoms, arranged in a repeating pyramid-like pattern. Actually it's a "tetrahedron," because a pyramid has a square base while this shape is triangular.

The reason diamond has this structure is because the carbon atoms all have a particular electrical charge, and the tetrahedron is the best shape for balancing that charge. The result is a material that's very hard; you can't squish it up even if you apply massive pressure across it evenly.

But if you take a long, thin diamond and start trying to bend it in half, the chemical bonds between the carbon atoms will try to resist this. Eventually, though, you will bend the diamond far enough that the crystal structure becomes misaligned. When that happens, the electrical charge between the carbon atoms is no longer perfectly balanced. There's a tipping point, and when that point is reached the bonds between some of the atoms snap and let go very suddenly.

The result is pretty spectacular: the diamond will shatter and spray tiny fragments everywhere. You'll be left with two diamonds with a ragged end on each one.

Natural diamonds often have faults inside where the bonds are weaker, and jewelers will use these faults to cut a round, pebble-shaped diamond into the faceted jewel we picture in an engagement ring. They do this by applying sudden force—the tapping of a hammer and chisel—to where they think the bonds are weakest. This results in a very clean and straight break. They can also polish the diamond by rubbing off just a few atoms at a time from the surface.

For all of these facts, the basic idea remains: crystals shatter because their atoms don't like moving inside their crystal lattice. There's no "give" in a crystal.

Metals are very different because of the way their bonds work. Instead of individual bonds between atoms, metals have what is called a "metallic bond" where there's a sort of sea of electrons surging around each atom. This means a crystal lattice of gold atoms can more easily balance out the stresses between bonds if you bend it. Gold is extremely soft and bendable and can be stretched into a wire, too.

Chemists call this bendiness versus brittleness "malleability" and the stretchiness "ductility." Some metals are very malleable, but if you try to stretch them, they separate. It's all about the bonds and the electrical energy between them.

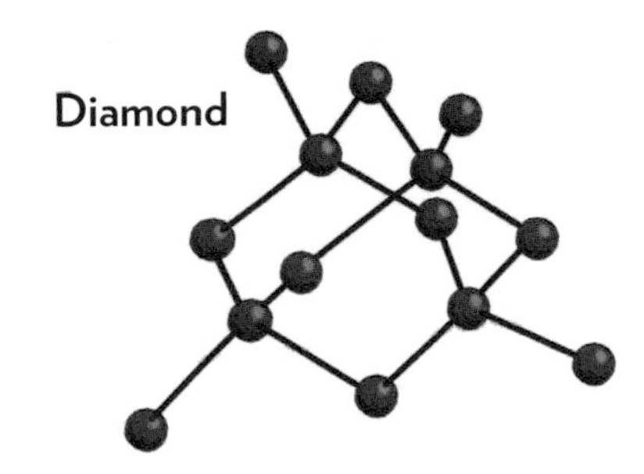

A diamond's crystal structure showing how each atom is bonded to four others, making it very strong but not very stretchy or bendable.

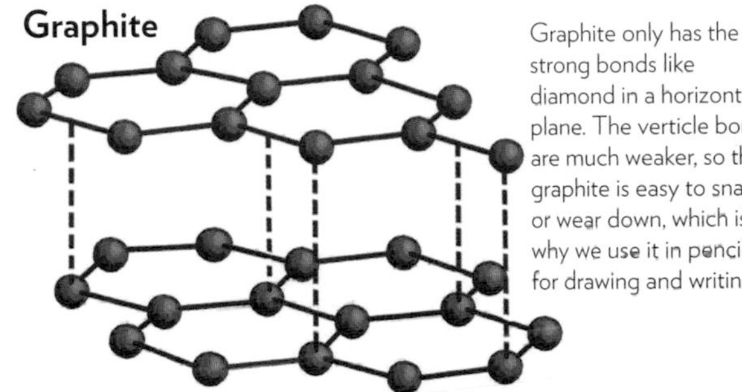

Graphite only has the strong bonds like diamond in a horizontal plane. The verticle bonds are much weaker, so the graphite is easy to snap or wear down, which is why we use it in pencils for drawing and writing.

Three ways materials can break

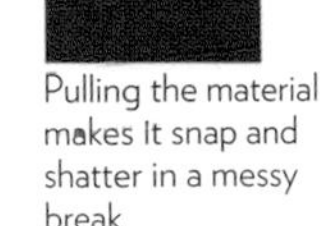

The material stretches and gets thinner until it's too thin and snaps

Pulling the material makes it snap and shatter in a messy break

The material is brittle but snaps much more cleanly

Q:

Why do soap and hot water make it easier to clean things?

While it's possible to wash clothes or dishes in plain cold water, using hot water and soap makes the job much easier. What's the chemistry behind this?

A:

Most of the things we consider dirt on our clothes and dishes are actually types of fat and oil that don't react with water. Soap helps them react. But heat plays an even more important role ....

Water is the essential ingredient in washing. It's an extremely versatile "solvent"—a chemical with just the right properties to allow other chemicals to dissolve into it.

But there is still a large number of chemicals that won't dissolve easily in water, and among these are some we encounter every day: oils and fats.

If you put on a white shirt and then roll around in some very dry, clean red sand, you'll find it pretty easy to just rinse out the shirt in cold water. The sand, made mostly of finely ground silicate rock, dissolves easily in water and gets carried away from your shirt if you swish it around in a bucket.

Dribble a serving of greasy fried chicken or gravy down the front of that same shirt, and you'll need to break out the soap and hot water. Not only because the grease itself stains your shirt, but because the grease will make regular old dirt stick faster—oil and fat attract and bind dirt more strongly than the bare fibers of cotton or polyester in your shirt.

We call soap an "emulsifier," which means it can help blend two liquids that normally don't mix—in this case oil and water.

But soap is actually a byproduct of a chemical reaction performed on a fat, called a fatty acid. Soap takes the form of a long chain of hydrocarbons, and one end of the chain reacts with water while the other end reacts with oils and fats.

When you mix soap in hot water, oil gets pulled away from your clothes or dishes and ends up suspended in the water. Then you can flush it away.

Seems straightforward, but why does hot water work better than cold water? It's not an illusion: if you wash your dishes at less than 90°F you can end up with a scum or residue of greasy soapy nastiness on the surface of each plate. At higher temperatures, oils and fats become less "viscous"—they get thinner and flow more like water. Washing in hot water makes it easier to flush the soapy, oily residue away.

Heat has another important role to play, especially when it comes to washing your hands. Lathering properly with soap and then rinsing under really hot water removes and kills significantly more bacteria than just rinsing your hands under cold water.

The problem with having bacteria on your hands is that you will inevitably rub your mouth or your eye, and those bacteria will then have access to your body. There are a lot of illnesses you can only get if you put bacteria from your hands into your mouth.

Those bacteria usually come from other living things, including the digestive tracts of other humans. They can also come from contaminated meat, a common source of which is your kitchen. Bacteria breed on bits of raw or old meat left during food preparation. Even when the area *looks* clean, it can be teeming with billions of bacteria. If you touch a contaminated surface and then touch your mouth, the bacteria can get into your body and infection begins.

Unless of course you wash your hands in hot, soapy water first.

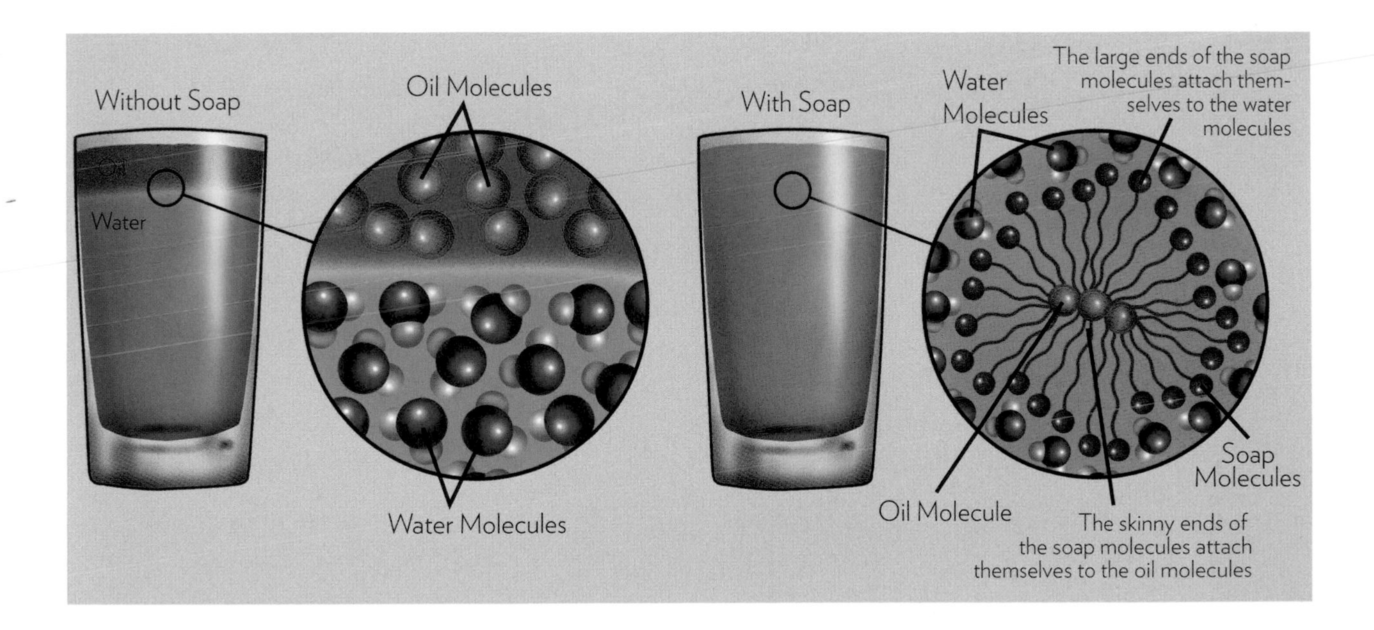

Q:

Why doesn't stainless steel get rusty?

Stainless steel is so called because it doesn't stain or rust; but how is this possible when normal steel rusts so easily?

A:

Actually, stainless steel is already rusty—it's just that the rust layer is incredibly thin and invisible! And stainless steel isn't the only metal that does this trick ....

When humans discovered how to process certain minerals into pure metals and make incredibly strong and beautiful structures, they also ran into a problem.

After several years, many of these structure and artifacts changed color, went all bumpy, and some even flaked apart and disintegrated.

This is a process called rusting, where the metal reacts with oxygen in the air. As the rust or "oxidization" reaction continues, the surface area of the rusty part increases, and that speeds up the reaction. It also lets rust push inside the interior of the metal object, ruining it.

Iron is especially vulnerable to rust, as the pure metal is slowly converted into iron oxide—which is weak and crumbly.

After a few centuries, humans discovered that mixing iron with carbon produced a strong new metal called steel. Steel is very strong and not too heavy, but it still rusts.

Toward the end of the nineteenth century, in Sheffield in the United Kingdom, chemists experimenting on steel discovered that adding about 13 percent of an element called chromium seemed to make the steel immune to rust. They called this new metal "stainless steel," but just like normal steel it does react with oxygen. The way it does this is very different, though.

Normal steel gets a layer of "iron oxide" on areas that are exposed to air and water. And this rust slowly eats its way into the metal object.

Stainless steel also reacts with air, but instead of iron oxide it gets a layer of chromium oxide. This molecule is crucially different, because once a layer forms, the oxidization layer stops. This "rust" doesn't penetrate the metal object (such as a sword) and weaken it.

Chemists don't call this rusting, they call it "passivation." And it's so fast and effective that even if you scratch a piece of stainless steel, the chromium oxide layer will form over the scratch almost instantly.

It is possible for stainless steel to rust, though. If it's used in something industrial, like a pump, and rubs against another surface, enough of the outer layer gets constantly scratched off that water and air can get in and react with the iron in the steel. When you open up your pump for servicing, the parts inside can appear bright red! Engineers call this "rouging."

Other metals are similarly rust-resistant. Aluminum doesn't rust very easily unless you bolt it to a different metal, whereby a whole bunch of other reactions can occur and cause corrosion. Titanium is very good at resisting rust and will go for years and years with no visible damage to its surface.

So why isn't all our steel stainless steel? Well, stainless steel is more expensive, and sometimes normal steel can be stronger. We can protect this steel with chemical treatments to the surface, or we can simply paint it! One of the most common treatments is called "galvanization," in which a protective layer of zinc is added to the surface of the steel via a chemical reaction. You can see this on boat trailers and street lamps—both of which are often exposed to water!

Rust and Stainless Steel

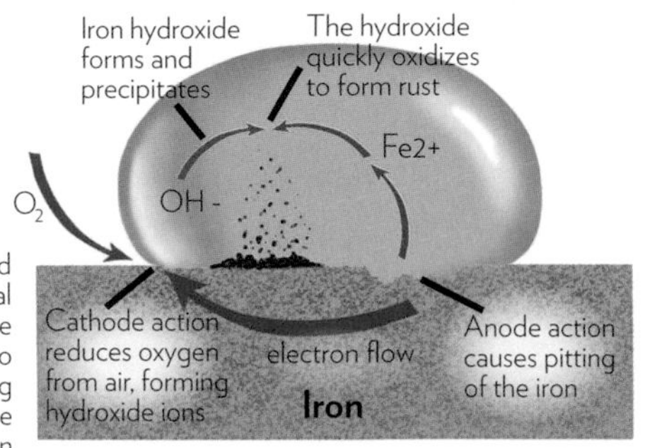

When iron rusts, and electrochemical reaction lets the rust eat further into the iron, exposing more of the metal to corrosion and forming pits

OH- = Hydroxide Fe2+ = Iron ion O_2 = Oxygen

Stainless steel forms a protective chromium oxide layer, which reforms even if it is damaged

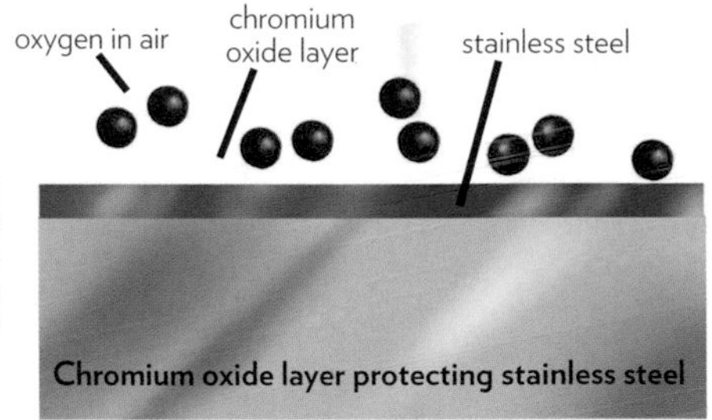

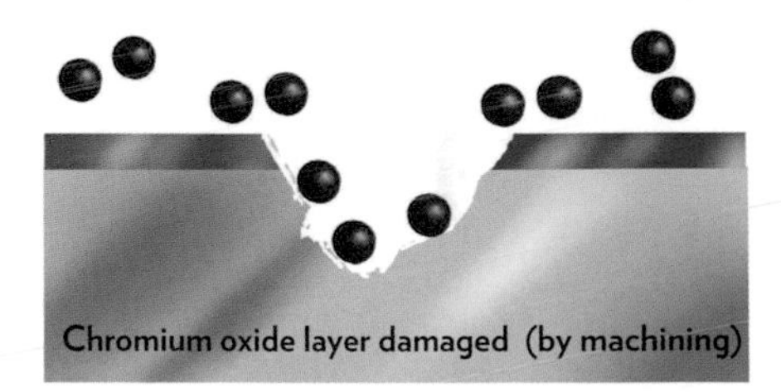

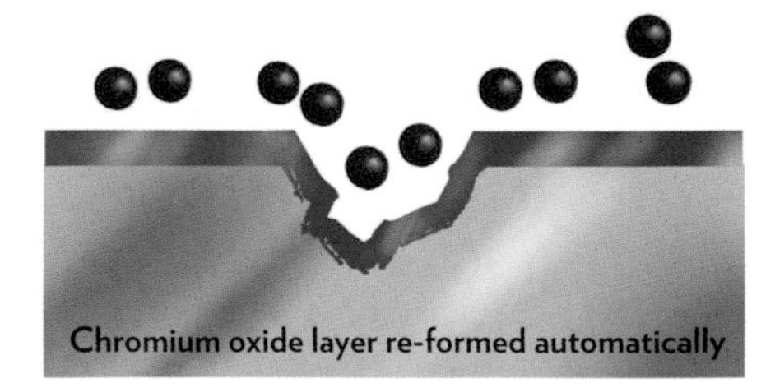

What gives gemstones their amazing colors?

The intense greens, reds, and yellows of gemstones make them desirable and valuable. But why are they so much prettier than, say, normal rocks?

A:

Gems are only colorful if they get contaminated by other materials, typically atoms of a metal. These impurities give them amazing color.

Humans have been mining the Earth for a couple of thousand years now, and in that time we've dug up a collection of rocks that we all agree are pretty enough and rare enough to be called "precious stones" or gemstones.

In Western cultures there are just four true precious stones: diamond, ruby, sapphire, and emerald. But in that list of four there are actually only three different minerals, because ruby and sapphire are both made of corundum. Diamond is pure carbon, and emerald is a mineral called beryl.

All four precious stones are made of common minerals, but it's the way these minerals are contaminated with other elements that makes them valuable—with the exception of diamond, which is also valuable because of the crystal structure that makes it very hard.

Emerald gets its distinct green color from trace amounts of chromium mixed into the crystal. Oddly enough, ruby also contains traces of chromium, but because it's made of a different mineral than emerald (corundum instead of beryl), it ends up a beautiful red color.

Sapphire is more varied and is pretty much defined as a gemstone made of corundum that *isn't* a ruby. Sapphires can be dark blue, purple, orange, or even greenish. Again, though, these colors come from chemical impurities such as iron, titanium, copper, or magnesium.

Diamond is a special case because, unlike the other colorful gemstones, we also think very pure, almost colorless diamond is valuable, too. But diamond can also have color, especially pink and yellow. While these colors can come from chemical impurities, they can also be caused by slight twists and kinks in the crystal structure of the diamond.

Gemstones and other crystals are basically giant collections of a chemical in which the individual molecules and atoms have arranged themselves into a surprisingly regular and geometric grid or "lattice." For instance, in diamond, each carbon atom is attached to four other carbon atoms in a sort of pyramid shape.

Light can shine through many crystals and make them appear clear or at least translucent. But if there are other minerals caught up in the crystal lattice, or the structure of the lattice isn't perfectly regular, light can be absorbed.

This is how gemstones get their color. When a beam of white light hits an emerald, the crystal absorbs all the light except the green, which it reflects back out of the crystal. Different trace minerals in different crystals reflect different colors of light.

Humans are so attracted to colored gemstones that not only do we pay big money for what are, chemically speaking, pebbles that are not superior to a lump of nice quartz—we actually don't think purer versions of the same minerals are as valuable!

There is a large number of other minerals we call "semiprecious," including topaz, opal, and lapis lazuli. These are more like colored rocks, without the intriguing crystal properties and amazing transparency of the true gemstones. But they, too, get their colors and swirling patterns from chemical impurities.

Mineral	Carbon	Corundum		Beryl
Gem	Diamond	Ruby	Sapphire	Emerald
Contaminant	None	Chromium	Iron & Titanium (blue)	Chromium
			Iron (pale yellow to green)	

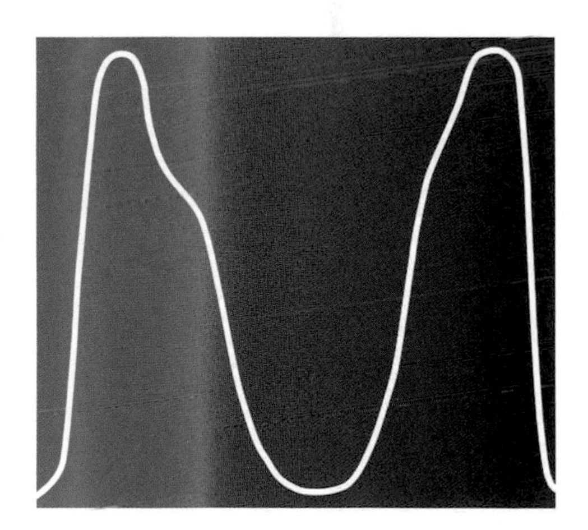

Q:

Why is frozen carbon dioxide called "dry ice"?

Frozen carbon dioxide is used in fire extinguishers and some fog machines. But they call it "dry ice." What's so dry about ice?

A:

Water ice turns into a liquid before it then evaporates into steam, but carbon dioxide skips the liquid part when it melts. So CO_2 ice never gets wet.

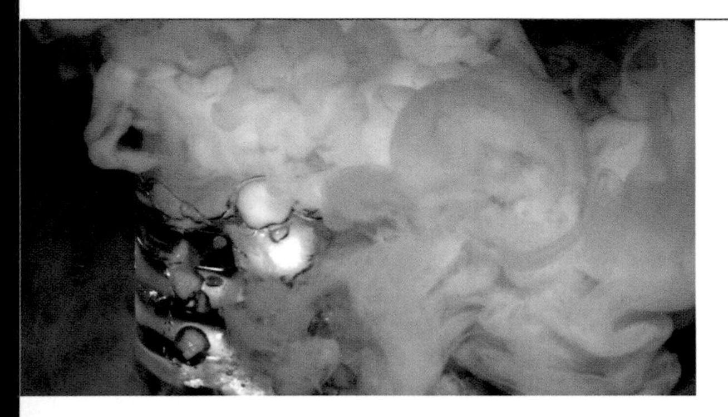

Under normal surface conditions on our planet, there are three so-called "states of matter"—gas, liquid, and solid. And the substance we see change states the most often is water.

Key to life on Earth is the way our planet is just hot enough, with just the right air pressure, for what's called the "triple point" of water. That means with just a little addition or subtraction of energy, we can make water a gas, a liquid, or a solid.

All solids can be melted, and you can think of any solid material as being "frozen." Water ice has some special chemical properties that make it different from a solid block of, say, iron, but the basic idea is the same.

If you heat iron to 2,800°F, it will melt into a liquid. If you keep heating it all the way up to 5,182°F, it will boil into a gas.

Carbon dioxide is the same. Under normal conditions here on Earth, CO_2 is a gas. If you chill it down to -109°F, it will freeze into a white ice that looks quite similar to water ice.

But when it comes to melting CO_2, we discover there's more to melting and boiling a chemical than just its temperature. The air pressure around the chemical is also very important.

People who live high up in the mountains already know their tea boils a couple degrees lower than that of people who live by the ocean. That's because the air is thinner at high altitude, and water boils at a lower temperature. In a similar way, carbon dioxide, unless it's kept frozen, will boil into a gas if the air pressure around it is less than five atmospheres—that is, five times the air pressure at sea level.

Since there's nowhere on Earth with air pressure that high except in special chambers and labs, any time carbon dioxide ice melts it skips the liquid phase and boils straight into a gas.

When you look at a block of dry ice melting, all the fog you see is just water in the atmosphere condensing against the very cold CO_2 gas. The CO_2 itself is invisible. When the dry ice melts completely away, there's no puddle or residue left behind. Thus the name: dry ice!

Dry ice is very useful because it's much colder than water ice. It's especially useful in insulated containers because it can keep water frozen without needing an external power source. We also use it in fire extinguishers because, as pure CO_2, it can smother a fire—which needs oxygen to burn.

The way CO_2 turns straight into gas from solid is called "sublimation." And it highlights why it's so important that Earth's temperature and air pressure be at the triple point of water. If our air pressure was very low, water ice would be like dry ice: it would boil into steam without forming a liquid first. Without liquid water, many of the chemical reactions in our bodies and the bodies of all living things wouldn't work.

Q:

What's so special about carbon, anyway?

Carbon, carbon, carbon. It's all you hear about these days. Carbon economy, carbon emissions ... what's so great about this particular element?

A:

Carbon is the foundation on which all life is built. Without carbon, there may not be any life at all. Yet carbon could end up killing us all ....

Life is made of chemicals, and life depends on chemicals. Without two important molecules—oxygen and water—nothing on Earth that we know about could survive. But there's a chemical even more fundamental to life than oxygen and water: carbon.

Even though oxygen is essential to make our energy, and water is essential to keep our cells working, none of this would happen without so-called organic compounds to carry the energy and use the water. And these compounds all have long chains of carbon atoms in them.

If oxygen is the walls of the house of life, and water is the roof, then carbon is the foundation. And also the mortar between the bricks. And all the furniture.

Carbon's chemical superpower is its ability to connect with up to four other atoms at a time. Not only can it make four connections, it requires relatively small amounts of energy to make it give up these atoms and break the chemical bonds that keep them attached.

Because of this, carbon can be part of millions of different chemical compounds. Think about it: there's carbon in the molecules that make up your eyelashes, but carbon also forms diamond—one of the hardest natural substances. Carbon floats around in the air as carbon dioxide, and it also makes up the wood of mighty trees. Wherever there's biology, there's a lot of carbon.

Life is a very chaotic kind of chemistry. Lots of different reactions happen in lots of different ways. The inner workings of the Sun, despite all that immense power and radiation, are much simpler than the set of chemical reactions needed to make a grasshopper jump.

Because carbon can be part of so many different kinds of reactions, it's the ideal basis of organic chemistry—the chemistry of living things.

But carbon's reactivity can also be a bad thing. For a start, it can bind up our atmospheric oxygen into carbon dioxide. That leaves less oxygen for us to breathe, but CO_2 also has the ability to trap energy from the Sun inside the atmosphere, and it can react with other chemicals in seawater to increase the acidity of the ocean.

Yet without carbon dioxide, our planet would be much colder and plants couldn't live. And the sea would become less acidic and eventually alkaline—and that would be bad for life as well.

Climate change scientists and governments prefer to talk about "carbon" rather than "carbon dioxide" because the carbon equation is more complex than just emissions of CO_2 from cars and industry. For instance, at the moment we rely quite heavily on fuels that are made up of carbon, whether that be oil, coal, or various plant-derived biofuels.

Even though these fuels are all quite different, to a chemist they're similar: long chains of carbon atoms with a few other things (especially hydrogen) attached.

The movement of carbon around your body is as important as the movement of oxygen and water, and the same goes for the planet as a whole. Of all the most common chemicals in our daily lives, carbon has the most versatile and complex job. No wonder it's the focus of such intense scrutiny and scientific study.

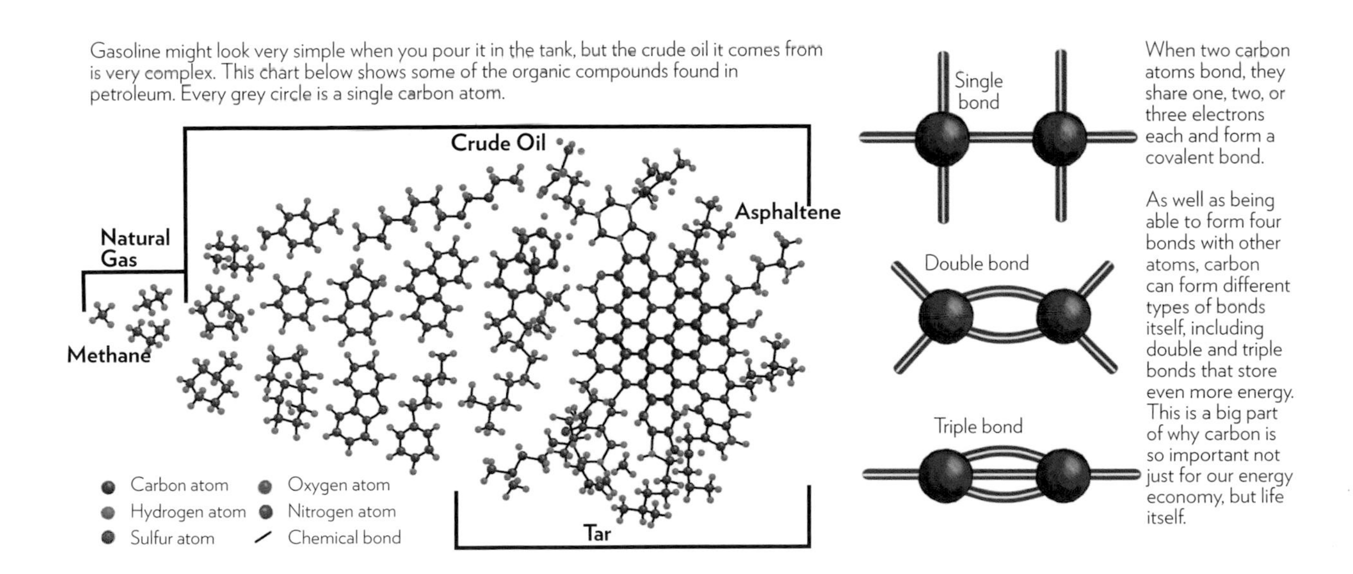

Gasoline might look very simple when you pour it in the tank, but the crude oil it comes from is very complex. This chart below shows some of the organic compounds found in petroleum. Every grey circle is a single carbon atom.

When two carbon atoms bond, they share one, two, or three electrons each and form a covalent bond.

As well as being able to form four bonds with other atoms, carbon can form different types of bonds itself, including double and triple bonds that store even more energy. This is a big part of why carbon is so important not just for our energy economy, but life itself.

Q:

How does a nonstick frying pan surface stick to the pan?

This question isn't as silly as you might think. How does *a surface that won't bond to anything bond to the rest of the object it's a part of?*

A:

Almost like Velcro does. Tiny pits in the metal surface of the frying pan trap other bits of the nonstick surface. It doesn't stick chemically, it sticks mechanically ....

One of the great inventions for bad cooks is the nonstick pan surface. There are a few different brand names for this surface, and some are considered better than others. But they all have two things in common: you can scratch them, and they can also poison you.

One popular nonstick surface is made of a compound called "polytetrafluoroethylene" or PTFE. This is a mix of carbon atoms and fluorine atoms in which the flourine grips on to the carbon so hard with such strong bonds that nothing else will react with or even stick to the surface. If you don't know how to fry an egg in a good-quality stainless-steel pan, PTFE is a great way to get lazy about washing up!

But PTFE is so effective at not sticking, manufacturers can't just paint it onto a steel pan. Instead, the steel surface is sandblasted so it becomes very rough and pitted. Then a liquid layer of PTFE is applied. In liquid form, it makes its way into all the little holes and bumps and cracks in the sandblasted surface. Then, after it dries, it's essentially trapped.

It's a little like pouring plaster into a mold and then being unable to pull the dry plaster straight out of the mold. The shape of the mold grips the plaster and keeps it from moving. It's not a chemical bond where atoms share electrons, just a simple mechanical grip.

Okay, so why don't these pan manufacturers just make the entire pan out of PTFE? Why have a steel base at all? Well, even leaving aside questions of expense and sturdiness and heat dispersion for the perfect steak, there's a BIG problem with PTFE.

Anyone who has done even a little chemistry will know that fluorine—part of PTFE's chemical makeup—can be very poisonous to humans if combined with other chemicals, such as might happen if it catches on fire.

In the case of PTFE, if you end up eating little flakes of the stuff, that probably won't do you that much harm because PTFE is inert. The same chemistry that stops your egg from sticking to it also stops it reacting with other chemicals in your body.

But if you accidentally leave your nonstick pan on high for a long time, the PTFE can get very hot. If it hits 572°F—which isn't impossible even on a domestic stove—the PTFE will break down and form something with the suitably scary name perfluorooctanoic acid, or PFOA.

This stuff is plenty poisonous, causing symptoms like tightness in the chest, coughing, nausea, and sweating. PFOA may also be carcinogenic, although studies are ongoing.

Used properly, your nonstick pan won't poison you, especially if you use wooden utensils and don't scratch it. But if the whole pan were made out of PTFE, parts of it would be right on top of the stove burner and be heated to very high temperatures every time you cooked something. This would produce lots of nasty gas and make the pan fall apart pretty quickly.

So PTFE coating it is. Still, for the safest gourmet fry-up, nothing beats stainless steel or cast iron. Sure, it means more washing up. But isn't that worth not getting poisoned?

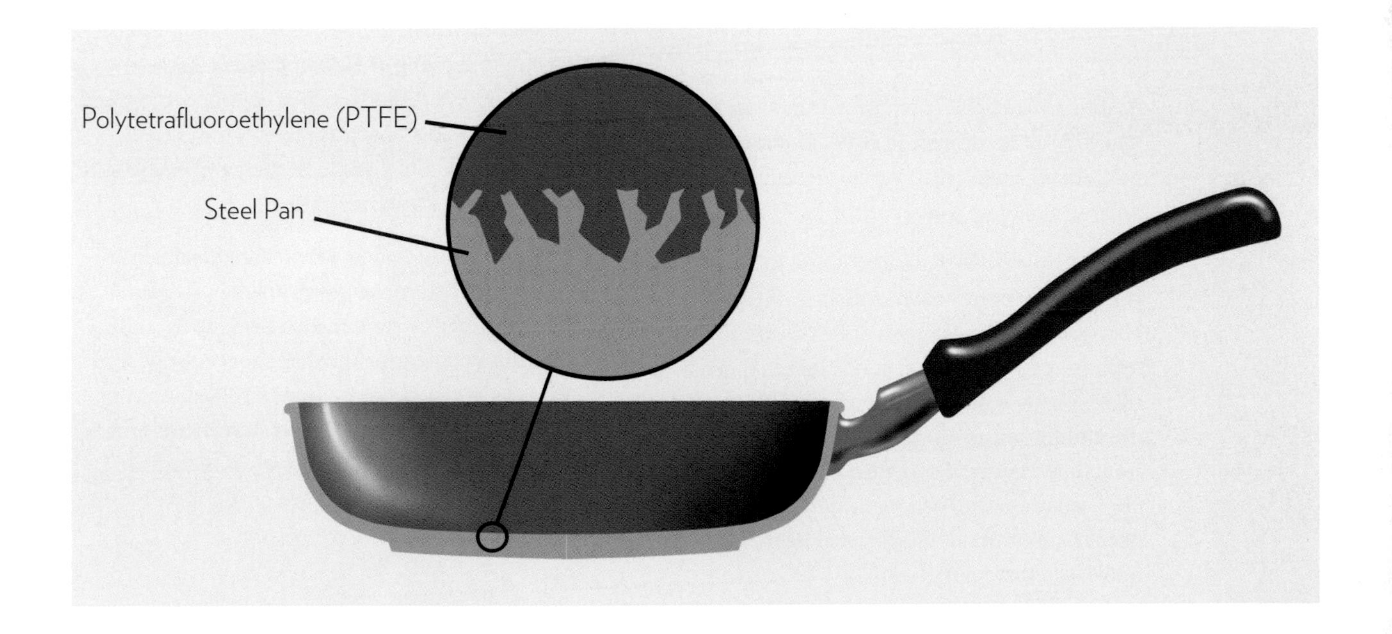

Q:

Why is life on Earth carbon-based?

We hear scientists talk about Earth life as being "carbon-based"—but what does this actually mean?

A:

It means every living thing on this planet is built from long chains of carbon atoms called hydrocarbons. As far as we know, it's the only kind of life possible ....

All life on Earth has a single common ancestor, if you go back far enough in time. About 3.5 billion years to be ... well, not precise, because we're still not certain about how life started.

The latest theories suggest that life might have begun deep in the ocean in places where natural chemical processes created rocky structures full of millions of microscopic pores. Inside these pores, increasingly complex molecules began to mix and eventually self-replicate. After even more evolution, these soups of so-called organic compounds chanced upon the "cell membrane"—a crucial development that let life escape its rocky cradle and move into the open ocean.

From there, the sky has been quite literally the limit for evolution. Life has evolved to fill every possible niche, and exploit almost every energy source on the planet, from simple sunlight to deadly acid.

But what every life form has in common is its basic chemistry. We're all made up of these long-chain carbon molecules. By "long-chain" we mean a string of carbon atoms all stuck together, with other kinds of elements attached around the edge. Because these compounds almost always include hydrogen and oxygen, they're called "hydrocarbons."

Chemically, hydrocarbons are ideal for life because they can break apart in reactions that release energy, and also form up again in reactions that store and transport energy. And life is, at its most basic, a set of self-sustaining chemical reactions that consume energy in one form and turn it into energy in another form.

This is possible because a single carbon atom has the ability to bond with four other atoms at the same time. There are other elements that have a similar ability—such as silicon—but the bonds these elements make take more energy to form and break. That means silicon-based life, if it existed on Earth, wouldn't be able to do as many reactions as fast as carbon-based life can.

Scientists think it might be possible for other types of life to exist on planets where conditions are different than they are here on Earth. Saturn's moon Titan is interesting because it has a dense atmosphere and lots of hydrocarbons on the surface. In fact, it might have whole lakes of alcohol! But it's very cold there—the average temperature is -290°F!—so any life would have evolved to move very slowly, maybe even slower than we can detect.

Studies of Titan show a puzzling lack of hydrogen in its atmosphere. This could possibly mean there is life there, living in lakes of liquid methane and ethane. This life would "breathe" hydrogen instead of oxygen and use ethane instead of water in its cells. This is all just supposition for now, until we can send probes with better instruments to find out for sure.

For now, the only life we know about is here on Earth, and carbon is an essential part of that life. Without carbon, we wouldn't exist!

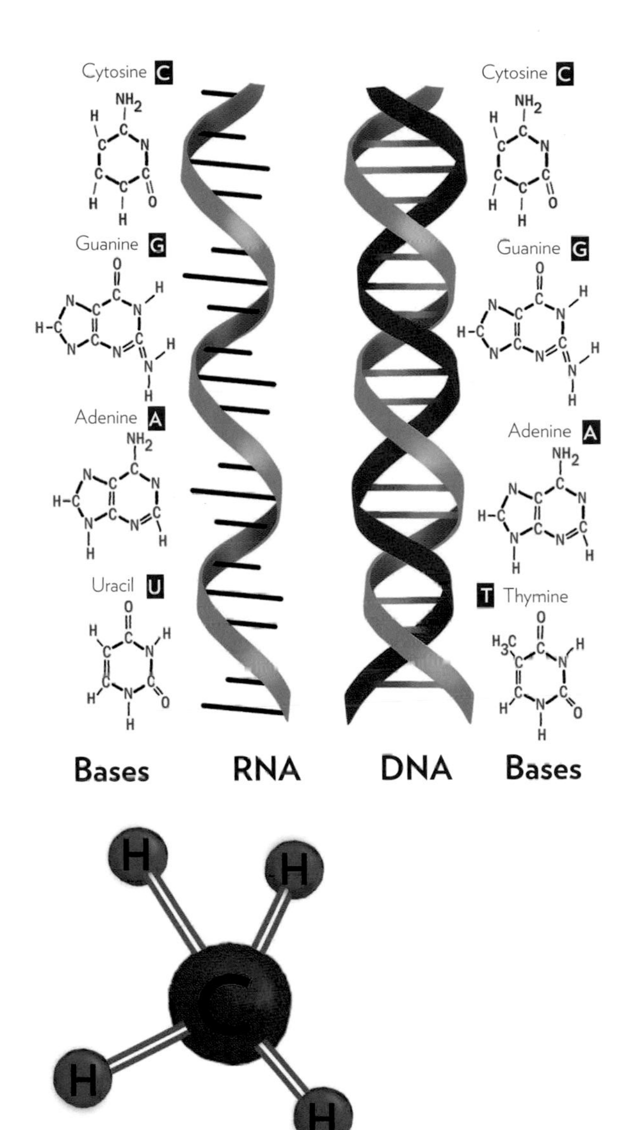

Q:

What exactly is an "organic compound"?

Scientists are always talking about organic compounds as being evidence of life. But what makes a compound organic?

A:

Life on Earth is based on a single chemistry that uses carbon, oxygen, hydrogen, nitrogen, phosphorus, and sulfur. We call that chemistry "organic" simply because it's the chemistry of life!

Chemists talk about different chemistries—distinct sets of molecules, compounds, and reactions that always seem to occur in groups. The way metals bond to each other and rust is one type of chemistry.

Another type is called "organic chemistry." This involves a huge number of different carbon-based compounds, water, and oxygen.

Today, organic compounds are those chemicals that contain a large amount of carbon. They're called this because for many centuries scientists thought these compounds could only be produced in living things, and that they would not otherwise occur in nature.

Modern chemistry and atomic theory have shown us that many of the organic compounds can be made by just mixing chemicals together in a lab. One of the first organic substances to be synthesized was urea—a component of animal urine. Because chemists were able to make it by mixing potassium cyanate and ammonium sulfate—neither of which is an organic compound—it changed the way we thought about chemistry forever. We realized that organic compounds are just one set in a broader system of chemistry and couldn't be rigidly defined after all.

"Organic" is a fairly sloppy term, scientifically speaking. Saying an organic compound is one that contains lots of carbon doesn't really help, because stainless steel has lots of carbon in it—and it's obviously not organic!

Any chemical that's produced by a plant or animal is definitely considered organic, though. Life on Earth has a tendency to produce very specific chemicals that wouldn't be found in nature otherwise. By searching for these chemicals, scientists can tell if a sample of soil or water has had life in it. The techniques developed here on Earth will be applied on other planets in the search for life—we're already sifting through the dust and soil of Mars looking for organic compounds.

It's tricky, though, because we've found so-called pre-organic chemicals in places like Saturn's moon Titan, but these *can* be produced without life. They're *almost* organic, related to life but not definitely made by life. It's a tantalizing hint that life might exist on other worlds, but for now we still don't know for sure.

Humans rely on other life forms to produce many important chemicals, because they would be much too expensive to make in a lab or factory by mixing raw materials.

One characteristic of organic compounds is that they have very long and complex chains of carbon atoms with other atoms attached. Making these in a lab is very difficult, and why should we spend all that time and energy when plants will make them for us?

The best example of this is sugar. Sugars are very complicated and essential for energy. Plants like corn or sugar cane will make more sugar than we can use, and it's delicious. Our attempts to make artificial sweeteners never seem to taste quite as good.

We rely heavily on plants because, as they assemble structures out of organic compounds at a microscopic level, they can build materials that are very strong and light. Wood from trees is an amazing substance that, treated properly, can be as strong as steel but much more flexible. And it's made entirely from long chains of carbon atoms!

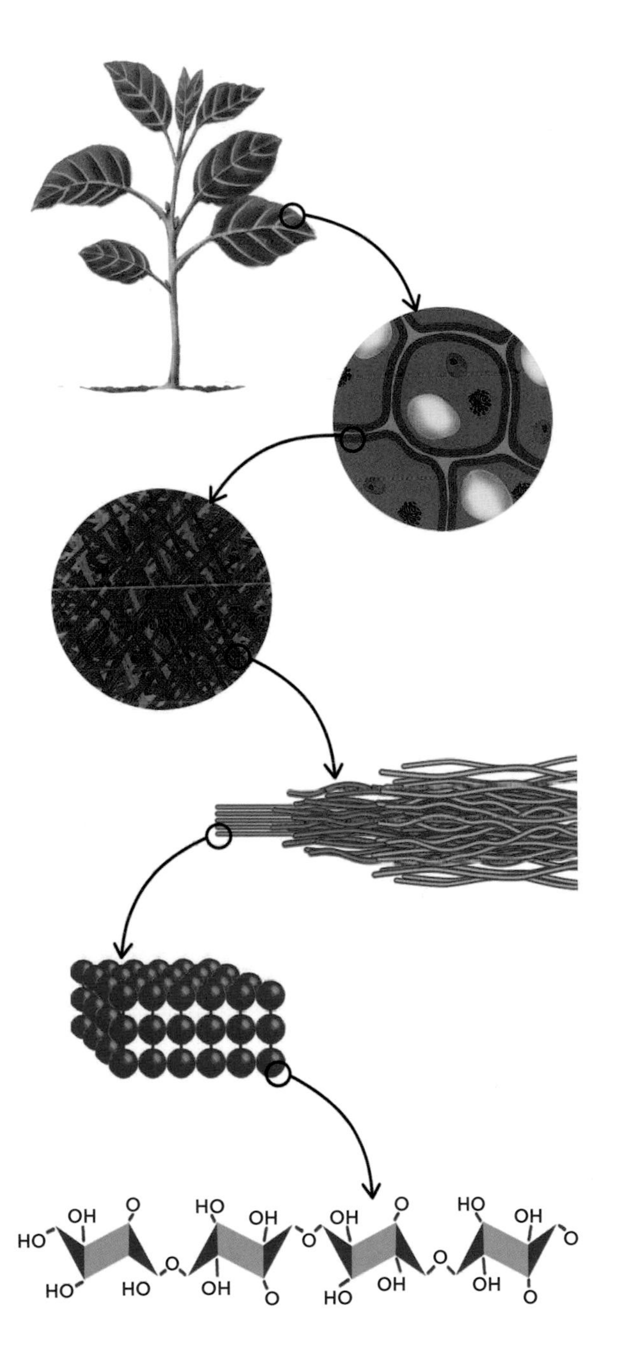

cosmology

Beyond the sky, the universe lies waiting for us. Cosmology is helping us take the first step ....

Less than a century ago, humans learned that the universe was much, much bigger than we'd ever suspected. Those lights in the sky weren't just other suns—some are even other galaxies made up of billions of stars in their own right.

We are tiny, an invisible speck in an invisible speck. There is more to nature than we can ever hope to explore. But we're still going to give it a red hot go!

Cosmology helps us understand our place in the universe. With powerful telescopes, we can search for other worlds. We can examine nearby planets, looking for clues to the origin and perhaps the ultimate fate of the Earth.

Most of all, we seek other life with which to share our stories and experiences. Will we find it anytime soon? Only the astronomers and cosmologists can answer that for sure ....

Q:

Why is the night sky dark?

In an infinite universe, shouldn't there be stars absolutely everywhere? In other words, the whole night sky should be stars, filling the sky in every direction. Why, then, is the night sky black?

A:

The universe might well be infinite, but it's not old enough for all the light from all the stars to have reached us yet. But even so, the night sky isn't as dark as you might think ....

Most of us just take the night sky for granted. The sun goes down, it gets dark, the sky turns black, and if there's not too much artificial light around us, we can see lots of stars.

But that's not the whole universe you can see up there. It's not even the whole galaxy. Humans can only see around 6,000 stars with the naked eye—and the Milky Way alone has 100 billion stars in it!

But if you add binoculars and telescopes, we can see many millions of stars. So part of the reason the sky isn't completely filled with stars is that our eyes aren't sensitive enough to see all the stars out there.

Even with our best optical instruments, there's still a lot more blackness than stars. And this is a surprisingly tricky problem for cosmologists.

If you accept that the universe is infinite (or very nearly infinite) in size, then that throws up some issues. Once you start messing with infinity, you have to admit that, with infinite options, there has to be a star shining no matter where you look. The night sky should, at the very least, be glowing with starlight bright enough for our telescopes to pick up.

The fact that it doesn't glow uniformly has made cosmologists stroke their chins for decades. Our current theories point to the darkness as evidence that while the universe might be infinite in size, it isn't infinite in age.

Across the bigger-than-immense distances between galaxies, light takes quite a long time to travel. Our nearest big galactic neighbor, Andromeda, is 2.5 million light years away. So light from Andromeda takes 2.5 million years to reach us. If Andromeda exploded or was swallowed by some kind of space monster in the time of, say, Julius Caesar, we won't be able to see that happen for another 2,497,943 years (as of 2013)!

The universe itself is about 13.7 billion years old. So any object that's farther than 13.7 billion light years away won't be visible to us, because its light hasn't had time—in the whole history of time itself—to reach Earth.

Actually it's more complicated than that, because the universe is expanding. We can see some objects farther away than 13.7 billion light years, because the space between us has stretched over time. But the principle remains the same: the light has to hit Earth for us to see the object. ("Actually it's more complicated than that" is a phrase you hear a lot in cosmology ....)

Our observations of the universe also show that matter clumps into denser regions with huge voids between them. All the "stuff" in the universe isn't spread out evenly, so there are gaps. Lots of very dark gaps.

The sky does actually "glow" everywhere, in every direction, though. After the Big Bang, the whole universe was very hot, mostly made of a sort of hydrogen plasma. It cooled, and matter clumped into galaxies, stars, and planets.

If we look deep enough into the sky with a sensitive enough radio telescope, we can still see this glow. Cosmologists call it the "cosmic microwave background," and it's good evidence that the Big Bang really happened.

Our eyes only see the small band of visible light. More importantly, because the universe is not infinitely old, light from every part of it has not reached us yet. We can only see the stars closest to us and the very oldest distant objects.

Q:

Is there anything in the universe bigger than a galaxy?

Asteroid, planet, star, galaxy—that's the hierarchy of stuff in the universe, with a few things like comets and black holes thrown in. But is there any structure bigger than a galaxy? If so, what does it look like?

A:

The size and structure of things in the universe are defined by gravity. And there are indeed bigger structures made of galaxies—they're just so big it's hard to see them ....

In the 1920s, astronomers discovered the Milky Way wasn't the only galaxy. Sensitive new telescopes took amazing images of distant spiral galaxies, elliptical galaxies, globular clusters, and other weird and wonderful forms.

So when it comes to "stuff," are galaxies as big as it gets? Is the universe just a random collection of evenly distributed galaxies? Billions of them?

For a long time, we thought this was the case. Part of the problem is that it's very difficult to figure out exactly how far away a galaxy is. Cosmologists know that at these sorts of scales, the gravity of stars and other galaxies can "bend" the light coming from very distant objects and give us the wrong idea about exactly where they are.

Imagine someone on a distant hilltop using a curved mirror to fool you into thinking they're standing several feet away from where they actually are. That's the sort of thing cosmologists have to correct for.

Eventually, with the help of supercomputers and an awful lot of math, we came up with a pretty decent image of the whole visible universe. And we've discovered that galaxies are grouped into even larger structures called superclusters, sheets, walls, and filaments. Between these are huge voids, vast spaces where there's pretty much nothing at all, just a faint wisp of hydrogen gas.

Because cosmologists usually stay up really late, many of these structures have cool names. There's the Sloan Great Wall, a sheet of galaxies about 1.37 billion light years long. There's the Eridanus supervoid. And there's a really large group of quasars called, uh, the Huge Large Quasar Group. That one is a massive four billion light years across and is the largest structure we know of.

Another neat term used by cosmologists is the "finger of God" effect. Because of the way the universe is expanding, if you don't correct for this in your observations, it can seem that in any direction you look there's a big chain of galaxies all pointed directly at Earth. But it's an illusion.

Our galaxy is part of a structure called the Local Group. It includes Andromeda, the Triangulum galaxy, and the Large and Small Magellanic Clouds. Plus around 50 other dwarf galaxies.

Our Local Group is part of the Virgo Cluster, which has somewhere between 1,300 and 2,000 galaxies in it. And the Virgo Cluster is part of the Virgo Supercluster, which includes over 100 other galaxy clusters—that's hundreds of thousands of individual galaxies. And *that* structure is part of the Pisces-Cetus Supercluster Complex.

How big is the Supercluster Complex? Well, it's a billion light years long and 150 million light years wide. And our supercluster makes up only 0.1 percent of its total mass.

If your brain isn't hurting right now, then just let us point out there are millions of superclusters in the universe. The good news is that at scales above supercluster complexes, filaments, sheets, etc., the universe looks pretty much featureless.

So if you can get your head around a string of galaxies a billion light years long, you don't need to worry about anything bigger.

Q:

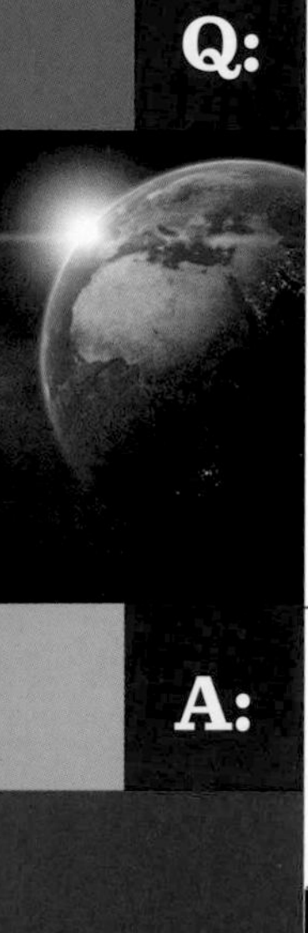

A:

How do we know how old the universe is?

At the moment, cosmologists believe the universe is about 13.7 billion years old. How did they come up with that figure, and how can they be sure it's accurate?

It's an educated guess. A very, very educated guess that can be backed up with sophisticated math and lots of observations. But the more data we gather, the more elusive a precise answer becomes ....

Figuring out a scientifically rigorous age (instead of just saying "about 13 billion years") for the universe is no small task, and organizations like NASA have spent millions of dollars and shot rockets into space trying to get a useful answer.

By "scientifically rigorous," we mean an answer that can stand up to some kind of scrutiny. Part of the problem is we don't yet fully understand how the universe is shaped, precisely how it began, or how big it is. Cosmologists do broadly agree on some things, though: the universe is fairly "flat," it had a beginning, and it's at least 92 billion light years across—and probably much, much bigger.

There's also general agreement that the age of the universe is 13.78 billion years, because observations of physical objects like galaxies and the amount of hydrogen in space have been fed into a mathematical model and that's the number that comes out.

But the number does rely on us having good observations. It's only "correct" if the observations we've made are complete and accurate. We think they are, but the universe has a habit of throwing curveballs just when we think we've got it figured out.

One of the easiest things to measure is the age of radioactive elements, like uranium, based on how much of a sample has decayed. From this, we can look at a chunk of radioactive material in the Solar System and say, well, that chunk is four billion years old, so the Solar System must be at least four billion years old. It could be older, because there could be evidence we haven't found yet—but it can't be younger.

The way cosmologists apply this kind of logic to the universe itself is very complex, but it's the same basic idea. By measuring the movement of galaxies, and also the properties of the "cosmic microwave background," cosmologists come up with a minimum possible age for the universe of 13.78 billion years. Give or take a few hundred thousand years.

If we're talking scientifically, the 13.78 billion figure takes us back to a time when the universe had just expanded from the Big Bang and everything was a sort of white-hot soup of plasma. It's the afterglow of this soup that makes the cosmic microwave background that cosmologists use to make estimates about the size, age, and structure of the universe.

At the moment, we can't "see" anything of the time that came before this white-hot soup. It's hard to know for sure how long it lasted, though there are a whole bunch of mathematical models that can give us a good idea.

NASA's Wilkinson Microwave Anisotropy Probe was launched into orbit in 2001, and for nine years it took measurements to help refine our understanding of the universe. It made some odd observations, such as the existence of a huge "cold spot" that our current model can't really explain. That's the problem with cosmology: the more observations we make, the more complex even the simplest questions become.

And that means the answer to "How old is the universe?" will almost definitely change as we learn more in the years ahead.

Discoveries of the Hubble Space Telescope

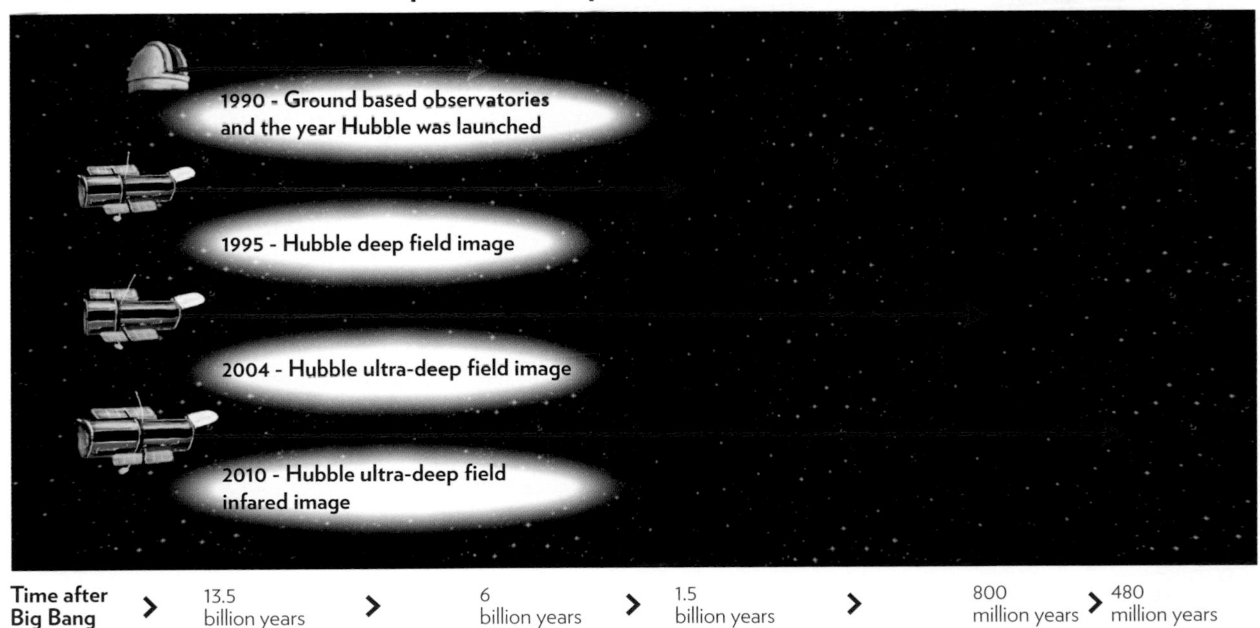

Q:

Why can't we see the bright center of our galaxy?

Artists' impressions of the Milky Way galaxy, seen from outside, show a spiral-shaped structure with a huge glowing center. But surely we should we able to see that center from Earth.

A:

The galactic center of the Milky Way is perfectly visible, just not to human eyes. With the right kind of instruments, you can spot it easily. But it's what is inside that's really freaky ....

We live inside what's called a barred spiral galaxy. A huge collection of stars, nebulae, and dust clouds all rotating around a central point.

Once we thought of this center as being shaped like a ball, but in the 1990s we discovered it's actually more like a gigantic peanut. The center is much longer than it is wide, making the shape of a huge, glowing bar. It's the brightest part of our galaxy.

We can't see this directly from Earth because of where our Solar System is positioned. There's a lot of gas and dust between us and the center.

If you know what you're looking for, on a very dark night when the band of the Milky Way is high in the sky, you can see extra brightness in the constellation of Sagittarius. It's sort of like holding your hand up to block a lamp on the other side of the room—you can't see the light, but you can see a glow around the edges of your fingers.

The dust and gas in the spiral arms of our galaxy block most visible light coming from the galactic center. But the dust doesn't block infrared light or X-rays. So we can use radiotelescopes, which "see" this kind of radiation, to learn about the structure of our galaxy.

While it's hard to get a precise distance, it seems the Earth is about 28,000 light years from the galactic center. If you find that hard to visualize, imagine the whole Solar System was the size of a quarter. Then the galaxy would be about 1,200 miles (1,931km) across, and we'd be 300 to 400 miles (483 to 644km) from the center. It's really big!

One mystery is that right where the galactic center is supposed to be, there's a very intense radio source called Sagittarius A* (yes, with an asterisk!). It's quite small, and astronomers can see matter rotating around it.

It's very likely that this radio source is in fact a gigantic black hole. Don't worry, though—it's not sucking us in, and most normal galaxies have a large black hole in the center.

Exactly how our galaxy is shaped is still something astronomers argue about. We can look at other galaxies far off in space and make assumptions, but it's extremely difficult to figure out exactly what our galaxy looks like from inside it!

Science-fiction writers often depict the galaxy as having two big spiral arms that wrap around themselves, but we could have many more arms than that, and they might not all be connected up neatly.

Today, there's general agreement that the Milky Way has four main spiral arms, named after the constellations we see them in, and a couple smaller arms that aren't joined to the galactic center, called spurs. We live in a spur called the Orion-Cygnus Arm.

Astronomers figure all this out by making observations, plotting points on a map, and then almost literally connecting the dots. Are they accurate? Well, the details of our galaxy have changed quite a bit in even the last 20 years! Our models will become more refined as we go, but until we can send a probe out of our galaxy (which with current technology would take literally millions of years!) the debate will go on ....

Q:

Is Saturn the only planet with rings?

All the planets in the Solar System are spheres except for Saturn, which has a huge and beautiful ring system. Why does Saturn have this special feature, and is it unique?

A:

Saturn just has the biggest, most prominent rings. Our probes have revealed Jupiter, Neptune, and Uranus all have less spectacular rings of their own.

When Galileo built his telescope and began examining the seven known planets, he found something very odd about Saturn. Humans have known about Saturn for all of our history—after all, it's a bright object in the sky that moves against the stars—but no one could have guessed how unique it was.

In 1610, Galileo first saw what he described as a "triple planet," three bright objects moving together in a row. He was baffled. Forty-five years later, Christiaan Huygens built a telescope sensitive enough to show that Saturn was surrounded by a mysterious disc, unlike anything anyone could have imagined.

Over the next 350 years, more powerful telescopes revealed something amazing: Saturn is girdled by amazing and beautiful rings. Tens of thousands of miles wide, but only a few hundred miles thick, the rings are divided into dozens of different bands, and there are even tiny moons orbiting inside them. They're made of trillions upon trillions of particles of dust and ice that glitter in the Sun.

The rings set Saturn apart as the jewel of the Solar System, but they did get astronomers thinking. What was so special about Saturn that gave it rings? It's smaller than Jupiter, but bigger than Uranus and Neptune. Was it all just due to chance? A moon made of soft material and caught in Saturn's gravitational pull, torn apart and smeared around the planet over millions of years?

While the size and beauty of Saturn's rings are unique, the mechanisms that made them aren't. All four of the gas giants have rings of dust and ice.

While the rings of Jupiter and Neptune are very thin and faint (some of them can't even be seen in visible light—you need special instruments), Uranus has a ring system almost as prominent as Saturn's—much narrower, but still quite bright. It just doesn't get put on postcards!

Uranus is a strange planet because, unlike every other planet in the Solar System, it doesn't point its equator toward the sun. Instead, it points its South Pole. This means if you traveled out from Earth in a straight line to Uranus, the rings would appear to point almost straight up and down (Saturn's rings would be pointing more-or-less toward you, tilted only slightly, like you see in the pictures.)

Neptune's rings are odd, too. They're so thin that particles can clump together in "arcs," or incomplete rings. If you went to Neptune, you might see a huge curve just hanging in space. But there is a whole ring there, it's just that most of it is invisible.

It turns out the rings of the gas giants aren't like moons—the "stuff" in the rings doesn't stay there forever. While it can take millions of years for material to cycle through the rings, it does eventually fall into the planet and burn up. But there's enough dust and ice floating through the Solar System for the planets to replenish their ring systems. So if you examine any individual piece of the ring, you'll find it's made of relatively new material.

It's quite likely that as our search for planets outside the Solar System continues, ringed planets like Saturn will turn out to be quite common.

Q:

Why do the gas giant planets have so many moons?

The gas giants are like little mini solar systems in their own right, with dozens of moons apiece. Why did they end up with so many and the rocky planets so few?

A:

The huge gravity wells of the gas giants trapped smaller objects in orbits around them. And it's a good thing this happens, too. Without the gas giants acting as interceptors, we could be in a lot of trouble ....

They don't call them gas giants for nothing. Take Jupiter, for instance. Jupiter is 88,846 miles (142,984km) wide. It could swallow 1,300 Earths and still have space for a couple dozen more. But its day is just nine hours, 55 minutes long, and gravity is 2.5 times stronger than here on Earth. If you could fly a regular passenger airplane on Jupiter, at normal speeds it would take nearly three weeks to circumnavigate the planet.

There's no surface like on Earth. Instead, Jupiter is one big ball of gas, though the pressure is so high deep in its atmosphere that the gas will be more like a liquid. There might even be liquid hydrogen in the core. Oh yes, and it has 67 moons.

Saturn, Uranus, and Neptune are smaller, but still gigantic compared to the little rocky worlds of the inner Solar System. And all four gas giants have much higher gravity than Earth—and that's key to why they have so many moons.

As objects orbit the Sun, they pass close to each other and exert gravitational pull. Mostly, the pull is so weak that nothing much happens. Over millions of years, though, the biggest gravity sources—the gas giants—attract objects closer and closer, eventually pulling them into orbit around them.

Jupiter is actually so big that four of its moons—Io, Europa, Ganymede, and Callisto—are nearly as big as the rocky planets. In fact, Ganymede is bigger than Mercury (by diameter, at least—though it's lighter).

These four moons weren't captured by Jupiter, but rather they formed out of Jupiter's so-called "subnebula." This was a huge disk of dust and rock that originally surrounded Jupiter after it formed. Think of Saturn's rings, but much bigger and denser.

Having four gas giants in the outer Solar System may be why we have life here on Earth at all. The gas giants protect us from really big asteroids and comets by intercepting them before they get too close.

The Solar System is surrounded by a region full of potential life-killer comets and other objects. The Kuiper Belt, as it's known, starts at the orbit of Neptune and goes way out. Pluto is a Kuiper Belt object. There could be up to 200 times as much stuff in the Kuiper Belt as in the Asteroid Belt between Jupiter and Mars.

Occasionally, Kuiper Belt objects will collide or pass close to each other, and one object might end up hurtling inward toward the Sun. If it hit us on the way through, it would be game over for humanity!

The gas giants act as celestial goalkeepers, intercepting or even deflecting objects like comets. As recently as 1994, we were able to watch a large comet called Shoemaker-Levy 9 break apart and hit Jupiter. One of the holes it left was nearly 7,500 miles (12,070km) across. There have been at least three other impacts on Jupiter in the last decade, though we didn't see the actual hits—just the scars and fireballs left behind.

These comets were the unlucky ones. Other objects get spun into orbits that will last hundreds of thousands of years, joining Jupiter's huge retinue of moons.

Jupiter (at left) has more than 318 times the mass of Earth (at right), is 1321 times bigger, and has 2.4 times more gravity

Q:

Why is the Moon so large?

The Moon is only the fifth largest moon in the Solar System overall, but relative to its parent planet it's by far the biggest. The Moon is 27 percent the diameter of Earth. How did we end up with such a big moon?

A:

The moons of Mars are captured asteroids, the moons of the gas giants were captured or formed from disks of dust. Our Moon is probably the result of a whole other planet hitting Earth, billions of years ago ....

For the first few millennia of our history, humans took the size of the Moon for granted. In a cosmic coincidence, the Moon as seen from Earth looks almost exactly the same size as the Sun. It was, for most of us, just the light that shone in the sky at night.

The Moon plays a huge role in life on Earth and is responsible for the tides, for stabilizing our rotation so the temperature on the surface doesn't change too much, and many other things.

But the moons of the other planets in our Solar System are relatively small compared to ours. Mars has a couple asteroids for moons, lumpy potato-shaped rocks called Phobos and Deimos that are each only a few miles across.

The gas giants have dozens of moons, some of which are very large. Ganymede, Jupiter's biggest moon, is bigger than Mercury! But even Ganymede, at 3,273 miles (5,267km) wide, is tiny compared to Jupiter, which is 88,846 miles (142,984km) wide.

Now look at the Moon and Earth—our planet is 7,917 miles (12,741km) wide (give or take), and the Moon is 2,158 miles (3,473km) wide. That's more than two thirds the width of the United States, and more than a quarter of the width of the whole Earth.

The reason the Moon is so large compared to the Earth is all due to how it was formed. Evidence suggests the moons of other planets were either captured by their gravity (as with Mars and its tiny asteroid moons) or the moons formed from the same spinning disk of dust and rock as the parent planet.

Our Moon, on the other hand, was formed much more violently. The Apollo missions were able to bring back real moon rocks for testing, and scientists discovered something very interesting. These rocks showed the Moon is made out of more or less the same material as Earth, except it's missing a lot of metallic iron.

There is one theory that fits these observations, and that's the Earth getting hit by a whole other planet as big as Mars (4,212 miles (6,779km) wide). We've used supercomputers to model this impact, and show that not only could it form an object as big as the Moon, but that also this object would have less metallic iron in it—just like the Moon really does.

Scientists call the mysterious planet from the past Theia, and it could have hit Earth around 4.3 billion years ago, long before there was any life to be killed off by such a catastrophe.

The collision would have sent Theia's metallic core into the Earth's own core, and some of Theia's outer layers would have been ejected into orbit and formed the moon—maybe in less than a month!

What's more, the thickened crust of the far side of the Moon suggests there was once a second, much smaller moon following in the Moon's orbit that eventually "pancaked" into the surface.

It's amazing that this cosmic accident could have given us the Moon we rely on for so much of the cycle of life today.

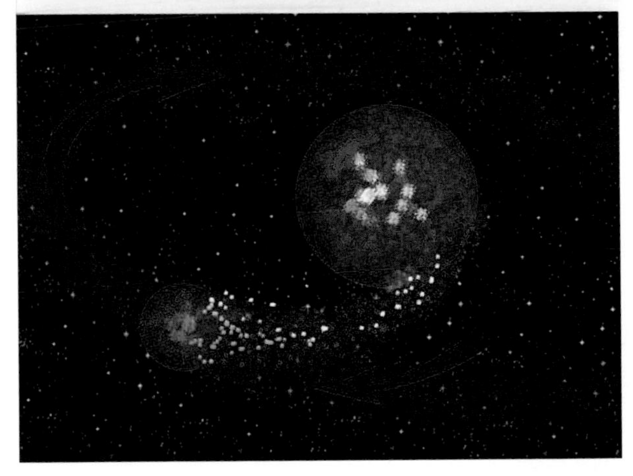

Q:

How do astronomers discover new planets?

The Holy Grail of astronomy today is to discover another Earth—but planets outside our Solar System are too small to see through a telescope. How do astronomers discover them?

A:

Extremely accurate measurements of the movement of distant stars can show a telltale wobble. Why does wobble equal planet? One word: gravity.

Almost all the light we see in the universe comes from stars. These balls of nuclear fire pour out massive quantities of light, enough to make clouds of gas glow and form beautiful nebulae.

Planets don't glow very brightly by themselves. We can only see them when they reflect sunlight. Even nearby worlds like Uranus and Neptune are extremely difficult to pick out unless you know exactly where to look.

Spotting a planet orbiting another star by looking for the light it reflects is, with current technology, impossible. Stars are just too far away and too bright for us to be able to see any smaller object near them.

But that hasn't stopped astronomers and planet-hunters from figuring out other ways to detect new worlds. One of the most important methods is to look for how a star is affected by the gravitational pull of its planet.

That's right: even though a star holds a planet in an orbit thanks to its massive gravitational pull, a planet's own gravity will, in a small way, pull back on the star.

For instance, as the Earth moves around the Sun in our orbit, we exert a tiny pull on our star, moving it toward us. It's a really tiny amount, but from a distance, someone could see the Sun shift slightly toward the Earth, then as the Earth travels around its orbit, shift slightly again in the opposite direction.

If you speed up this movement, it would be possible to see the Sun "wobble" back and forth in place as its planets travel around it. Now, the Earth is a relatively small world and has a small effect—we have been able to detect Earth-sized planets, but at the moment it's much easier to spot so-called "super Jupiters."

These massive gas giants make their stars wobble dramatically. The first planets we found outside the Solar System were super-Jupiters, huge worlds that orbit very close and very fast around their stars.

Over the last decade, our planet-hunting techniques have improved, and we've launched better satellites into orbit around Earth—a NASA telescope called Kepler is one of the most important.

Kepler's main job is to find planets, and its instruments are so sensitive it can even detect the shadow of a planet passing across the front of a star—like our Moon making a solar eclipse.

By seeing how the light from the star changes as the planet passes—or "transits"—scientists can even figure out what color the planet is.

There's one planet with the rather unromantic name of HD 189733b that is colored a bright blue, even bluer than Earth. It's not very Earth-like, though: it orbits close to its star, one side is permanently dark, and on the surface it rains molten glass.

Need something weirder? How about a planet so big it's almost a star? Or one with a surface made of diamond? Or a super-Earth covered in an endless ocean?

We've only surveyed a fraction of the sky for planets so far, but at the current rate of discovery, it's looking like there are hundreds of billions of worlds in our galaxy alone.

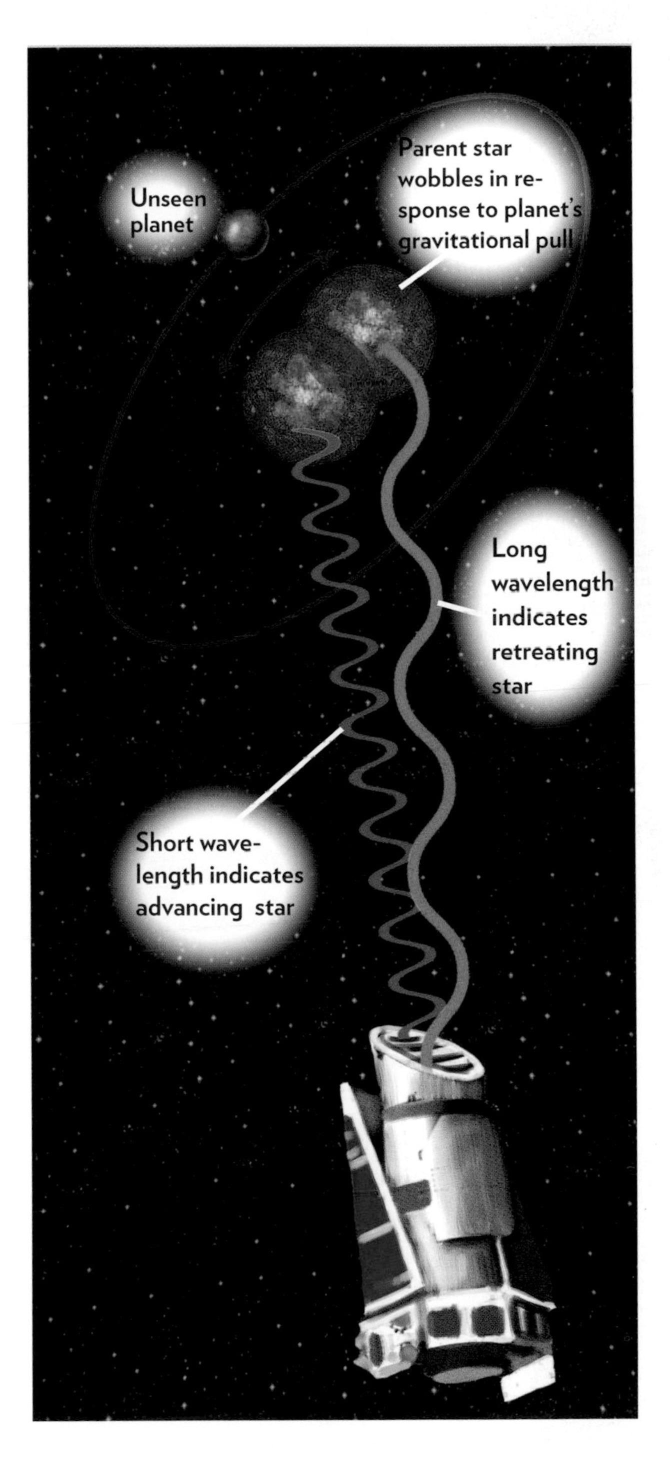

How much of the universe can I see with the naked eye?

On a really dark night, far from the city, the sky is absolutely crammed with stars. How much of the universe can we take in just lying on our backs in a field?

A:

The unaided human eye can see an infinitesimally small portion of the universe—barely 3,000 stars and a handful of other objects at a time. The whole picture is much bigger ....

Today, there are roughly 6,000 stars bright enough for us to see while standing on the surface of Earth. Add to those some of the gas clouds in the Milky Way, a handful of other nebulae that show up as pale smudges in the sky, the Small and Large Magellanic Clouds that are nearby galaxies, and the Andromeda galaxy if you know where to look.

Of those 6,000 stars, you can only ever see around half of them at a time because the horizon will block your view of the rest. If you wait patiently, the rotation of the Earth will bring more of them into view as the night passes.

Part of the reason we can only see 6,000 stars is because of light pollution. Even far away from cities, the atmosphere reflects enough light to wash out the faintest stars. Before industrialization the night sky was quite a bit darker, and humans may have been able to see as many as 45,000 stars—though because of the way the atmosphere absorbs starlight, it might have been fewer.

The brightest star you can see in the Northern Hemisphere is Sirius, the Dog Star. In the Southern Hemisphere, Alpha Centauri is both the brightest star and also the closest to Earth (technically the closest star is its smaller companion, Proxima Centauri, but they're so close they look like one star).

Stars are very faint compared to the normal things we look at, and our eyes aren't well adapted to naked-eye astronomy. We have to use the light-sensitive "rods" in our retinas rather than the color-sensitive "cones," so stars mostly look white or greyish. If you really concentrate you can pick out some stars that are redder or bluer than others, but it's tricky.

Starting with Galileo in the seventeenth century, humans developed telescopes to massively boost our ability to see the universe. And when we started to hit the limit on optical telescopes, we invented radio telescopes that allow us to "see" through gas clouds and pick out extremely faint and distant objects.

It was the astronomer Edwin Hubble in the late 1920s who first realized there were other galaxies, and over the last 100 years we've discovered the universe is much, much bigger than we thought.

The current estimate is that the universe has 100 billion galaxies, each with 100 billion stars in it. So the number of stars is ... bear with us ... 10,000,000,000,000,000,000,000. No, we don't have names for them all.

Meanwhile, the search for so-called exoplanets continues, and at the rate we're finding them in orbit around stars here in the Milky Way, it's likely there are many more planets in the universe than stars.

We've come a long way in our understanding of the universe from those long, dark nights lying on the hillside and tracing the shapes of mythical creatures, gods, and heroes in the patterns of stars overhead. It's likely we'll expand outward to colonize at least some of those stars. Who knows—the next time you look up at the night sky, you might see the sun of one of your future descendants.

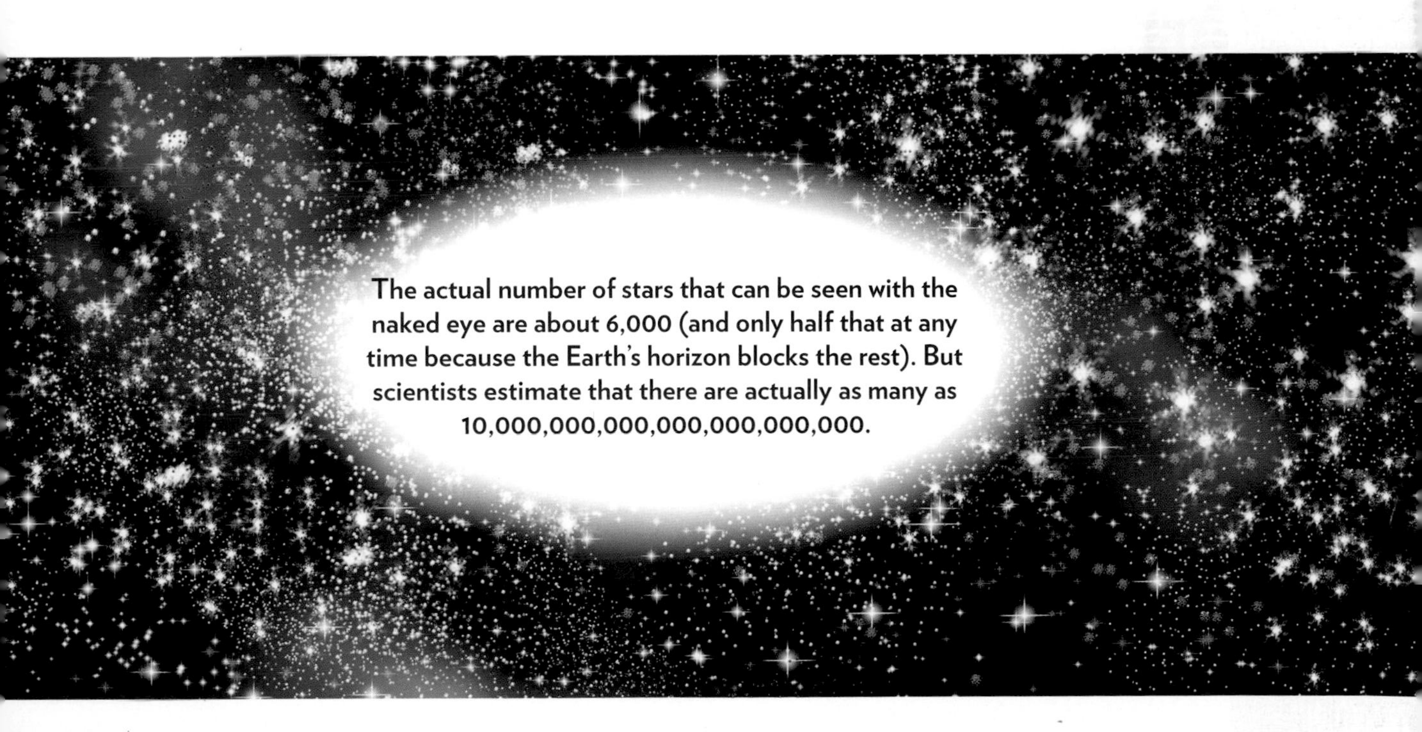

Q:

Why do we use "light year" as a measure of distance?

One of the more confusing concepts in cosmology is the way we measure distance between stars, because we use a word for measuring time. Light year is a weird term, so why do we use it?

A:

The distances between stars and galaxies are so vast that our Earth-bound measuring systems are far too small. The "light year" was invented so cosmologists could use smaller numbers! Except now they have a new, better measurement ....

Sometime around the eighteenth century, astronomers began to get serious about measuring the distances between various objects in space.

They'd already figured out how far the Earth is from the Sun (92.96 million miles or about 149,668,992km) and decided that was too big a number to have to write down all the time. So they came up with the "astronomical unit" or AU. The Earth is therefore 1 AU from the Sun. Jupiter is 5.2 AU from the Sun. Much easier than writing 483,370,198 miles (777,908,927km).

The next step was to measure the distance to a nearby star. In 1838, a German astronomer named Friedrich Bessel used a combination of complicated lenses and even more complicated math to figure out a star called 61 Cygni was 660,000 AU from the Sun.

Clearly, astronomers were about to run into the same problem they'd had with miles. Stars were millions of AU from the Sun, so a neater measure was needed.

At this time, scientists were starting to realize that light is the fastest-moving thing in the universe. So it made sense to use some property of light's speed to measure distance. Bessel decided that the distance light traveled in one year would be a useful measure.

And so the term "light year" was coined. Bessel said his star 61 Cygni was 10.3 light years from Earth. Today, we know it's 11.4 light years away, but Bessel didn't have any help from computers, so his calculation is impressively close.

Just so you know, a light year is about 6 trillion miles/9.5 trillion kilometers. That number is so huge it's practically meaningless. Think about it this way: the Moon is about a light-second away from Earth (the distance light takes one second to travel) and Earth is about eight light-minutes from the Sun. Our nearest neighbor star, Proxima Centauri, is four light years away. The galaxy is about 100,000 light years across. And our big galactic neighbor Andromeda is 2.5 million light years away.

But wait, we're not done here. Because this is cosmology we're talking about, nothing is ever simple. While a light year is a great way of talking about interstellar distances without filling up the page with numbers, it's hard to match light years with actual observations of stars from the surface of the Earth.

Toward the end of the nineteenth century, astronomers started using a different measurement that was more useful. As with most things in cosmology, explaining this new measure involves a lot of math and triangles and orbital speeds and things, but at the end of the day it has to do with how the position of a star appears to change in the sky based on where Earth is in our orbit.

The new measurement was named in 1913 by English astronomer Herbert Hall Turner. He called it a "parsec." You might have heard this word in a certain famous science-fiction movie, where a roguish space freighter captain claims he had made "the Kessel Run in less than twelve parsecs."

A parsec is about 3.26 light years, but it's better for cosmologists because it's more accurately defined than a light year. And understanding exactly how far objects are away from Earth is vital in building our picture of what the universe really looks like.

Astronomers need to measure very large distances. In popular science literature the light year is commonly used (1 light year = 5.878625 trillion miles or the distance light travels in one year). Scientists however prefer to use the parsec as a measure of long distances because it is more accurate and easier to calculate.

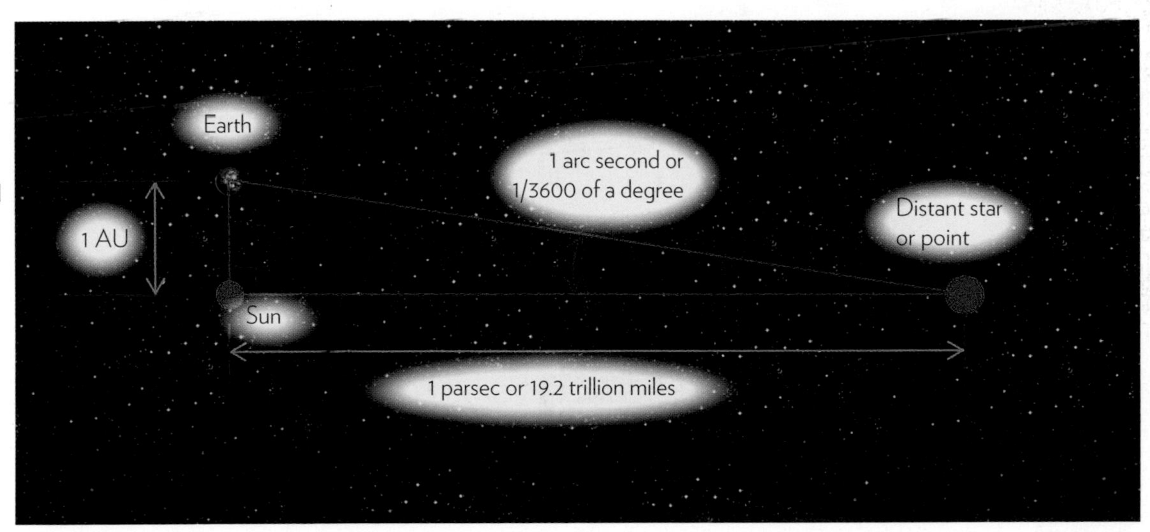

If the base of an imaginary, right angled triangle is the line from the Earth to the Sun and the other two sides intersect at an angle of 1 arc second, then the point where they intersect is one parsec from the right angle.

Q:

What makes the stars twinkle?

Twinkle, twinkle little star—it's one of the first nursery rhymes many of us hear. But if the stars are really suns like our own, huge balls of nuclear fire, why do they twinkle?

A:

Earth's atmosphere causes stars to twinkle as the air overhead moves. Wind patterns and differences in temperature are the main culprits, but the human eye has a role to play, too ....

While the whole of Earth's atmosphere is about 500 miles (805km) thick, the life-giving troposphere reaches up only 10 or so miles (16km) at the equator and contains 80 percent of all the oxygen, nitrogen, and other gasses.

The troposphere is thick enough that when starlight passes through it, the beams of light get bent and twisted by turbulence—simple wind.

The longer you look at a single star, the more chance the light you're seeing will be very slightly jinked around by air moving overhead. Stare at the star long enough, and you'll see it twinkle.

This twinkling is more obvious if you look at a star close to the horizon. The light from this star has to come through more air to reach your eye, so it has more chance of being bumped by turbulence.

So why doesn't the Moon twinkle? Its light has to get through the same amount of the atmosphere, after all. But because the Moon is such a relatively large object, your eyes and brain filter out tiny changes in the light hitting your retina.

Stars are so far away they appear as what's called a "point source" of light. The image of a star in the sky has no size, it's a tiny pinprick, so small that it only activates a single "rod" sensor in your retina.

When the atmosphere affects the star's light, the image of the star gets "bumped" across to a different rod in your eye, and your brain picks this up and interprets it as the light from the star flickering, or twinkling.

Because the Moon is big enough to activate lots of rods in your retina, the way the image twinkles is simply ignored by the brain. However, if you look at the Moon through binoculars or a small telescope, especially on a muggy summer's night, you can quite clearly see the surface shimmering. Depending on conditions, it can even look like the surface of the Moon is under a thin sheet of oily water. This is the Moon "twinkling."

Twinkling is kind of romantic when you're lying on a hillside, but if you're trying to take accurate measurements using an Earth-based telescope, it can be very frustrating. Atmospheric turbulence can ruin years of work!

Telescope engineers have come up with a range of different systems to correct for twinkling—the official word for the phenomena is "scintillation." By far the coolest one is "adaptive optics."

Computers monitor the light hitting the telescope's main reflecting mirror and make measurements of how much the atmosphere affects the image. Then the telescope uses motors to rapidly bend its mirror back and forth to correct for atmospheric turbulence, canceling out the oscillations and almost completely getting rid of twinkling! This technique has given us amazing new images of the universe that rival pictures taken from telescopes in space, like the Hubble.

Space telescopes outside the atmosphere still get the best pictures overall, though they're not immune to twinkling out there either. The particles coming from the Sun—the so-called solar wind—and the very thin mix of hydrogen and nitrogen that fills space can cause "interplanetary scintillation," or twinkling on a galactic scale!

Nothing's ever easy in cosmology. Not even taking a few pictures.

Q:

Why does the Milky Way glow?

The Moon reflects light from the Sun, and stars burn with their own light, but what makes the bright band of the Milky Way so luminous?

A:

The bright band is one of the arms of our galaxy, and while there are millions of stars inside it, gas is also heated until it glows. But that glowing band may not be as permanent as you think ....

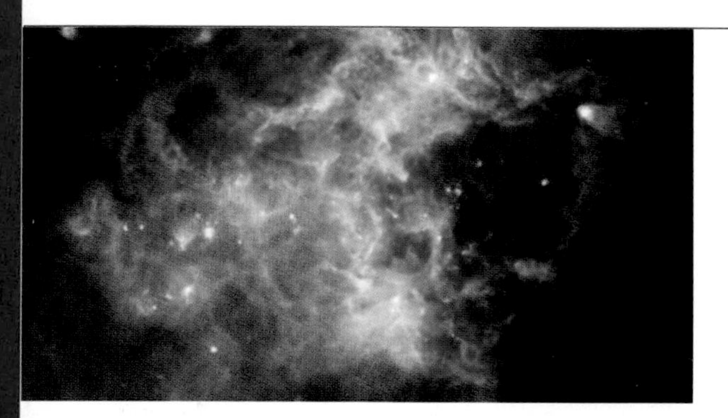

The Milky Way is the pale glowing band that stretches across the sky on very dark nights. Depending on the time of year, it may stay close to the horizon or pass nearly directly overhead. And if you're anywhere near an artificial light source or a bright Moon is up, it gets washed out. So it's much dimmer than the regular stars.

What you're looking at is one of the spiral arms of the Milky Way galaxy. Our galaxy has at least four arms, and the Solar System is in one of the minor arms or spurs, called the Orion-Cygnus Arm. The band across the sky is the arm "across" from us, toward the galactic center.

Of course, technically every star you can see is part of the Milky Way galaxy, but that understanding came many thousands of years after we started calling the band across the sky the "Milky Way."

Even though "Milky Way" is an English term, many cultures around the world named this feature after something to do with milk. In fact, the word "galaxy" comes from the Greek word for milk!

The milkiness we see is billions of stars. There are so many and they're so far away that our eyes can't make out individual points of light—they all just blur together into a milky glow.

If we use a powerful enough telescope, we can make out the individual stars and also huge regions of gas and dust. Some of the dust is thick enough to block starlight, which is why the Milky Way has a sort of mottled texture to it. The irregularities in it are regions of gas blocking starlight. Other gas has been heated up enough by other stars to glow and add more light to the whole arm.

These spiral arms aren't "structures" like, say, the rings of Saturn or the continents and oceans of Earth. They're more like the ripples you see on a pond, or the waves on the ocean. As the galaxy spins around, waves of energy pass across the surface, bunching stars and gas into these arms and then passing onward.

There are many other parts of the night sky that glow with milky light. The Large and Small Magellanic Clouds, for instance, which are smaller galaxies that orbit the Milky Way, a little like moons—but moons made of billions of stars each!

There are also many nebulae visible with the naked eye, such as the Orion Nebula, which is the middle "star" in Orion's Sword. If you look at it with binoculars you can see it's not a sharp point, but diffuse. It's a region of gas and newly created stars more than 24 light years across.

Our Sun won't always lie inside the Orion-Cygnus Arm. It takes a mind-blowing 220 million years for the Sun to make one orbit of the galaxy, and in that time it will move through regions with more dust and with less dust, and other stars will come close to us and then move farther away. The spiral arms will ripple and distort, combine and twist apart. Over the eons, the night sky will be as changeable as the surface of a river.

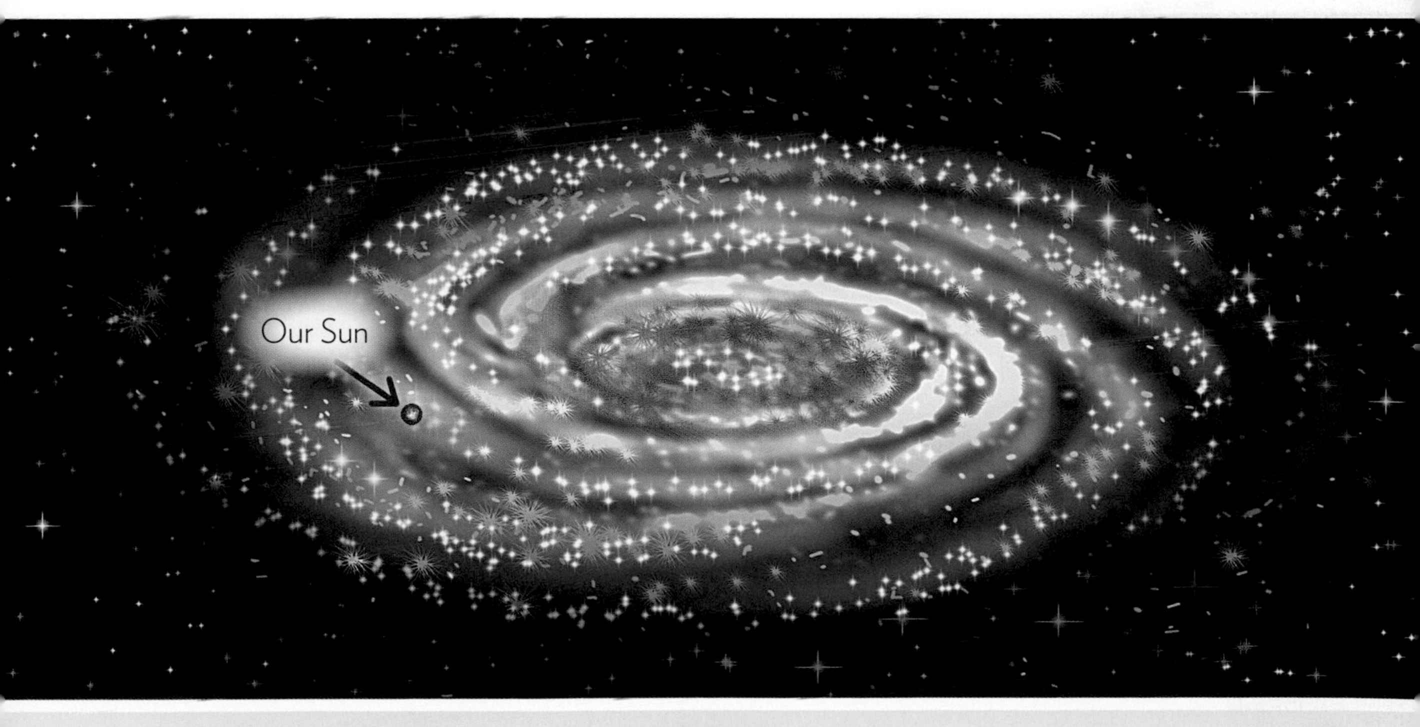

Q:

How do astronomers figure out how far away a star is?

We take it for granted that, say, Alpha Centauri is four light years from Earth, but how did we figure out that distance in the first place?

A:

The stars appear fixed in the sky, but with a good telescope you can see them move slightly as the Earth orbits the Sun. This movement, or "parallax," is the basis for a bunch of really complicated calculations ....

Working out how far things are away from you here on Earth is relatively simple. You can look at a building you know the height of, measure how high it appears to be from where you're standing, and then do some trigonometry to get the answer.

After you've done that, you can walk toward t he building and count out the distance, double-checking your calculations. Easy! Well ... as easy as trigonometry, anyway.

Astronomers use a similar technique to figure out how far away a star is, except there are two main problems: they don't know for sure how big the star is, and they can't double-check the distance by travelling there.

One of the problems is that to the naked eye, all stars appear to be exactly the same size, and that size is no size at all. Stars are "point source" lights, they have no visible height or width.

As we invented better telescopes, scientists figured out they could get a pretty good idea of a star's distance by taking into account how the position of the Earth in its orbit would slightly change the apparent position of the star in the sky. This difference would be most noticeable for observations taken six months apart—when the Earth was on the opposite side of its orbit from the first observation.

It's like setting a candle up 8 feet (2.6m) from you, then looking at it with just your left eye. Then close your left eye and look through your right. The candle will appear to move slightly to the left.

Astronomers call this "parallax," and they use it to measure the distance between the Earth and various stars. It takes quite a bit of patience! But for stars less than 100 or so light years away, it works really well.

We've also figured out that some stars have a very consistent light output. Astronomers call them "standard candles," and if a standard candle star looks less bright than it should, then that means it's farther away.

These measurement systems only really work for distances up to a few hundred light years. Farther out than that and it gets hard to build instruments sensitive enough to pick up the amount of parallax in a star's position.

There are many techniques used to figure out really huge distances—such as the gulf between the Milky Way and the Andromeda galaxy—but they all rely on finding certain types of stars and deciding how bright they look and what that means about their probable distance.

If it all sounds kind of vague, that's because it is. Astronomers call their distance calculation system the "cosmic distance ladder." Different "rungs" on the ladder give different certainties of measurement. The lowest rung is the system we use to calculate real distances, like how far we are from the Sun or from Venus. It's pretty reliable. The higher up the ladder you go, the bigger the distances and the more uncertainty in the answer.

Every few decades, cosmologists figure out a new system to further refine their answers. And the general rule so far has been this: the universe is always bigger than we think.

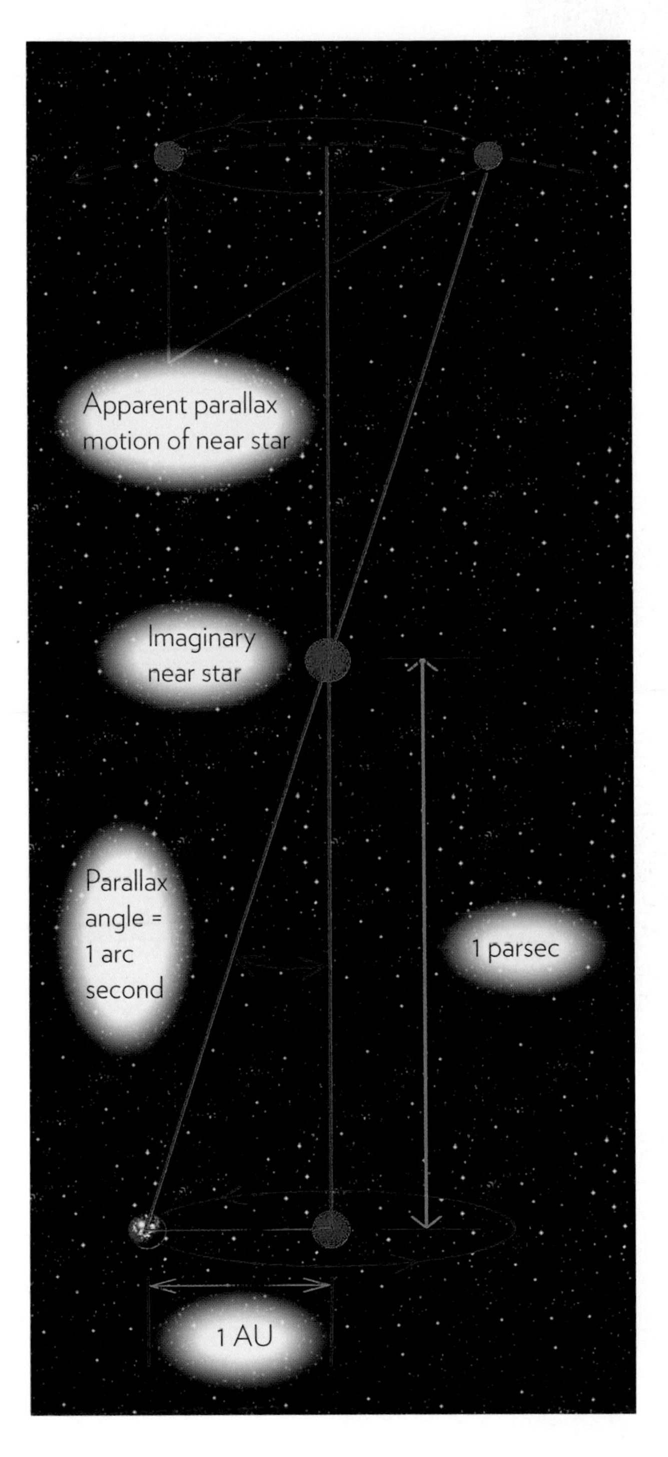

Q:

Why doesn't the North Star move in the night sky?

All the stars slowly cross the sky during the night as the Earth rotates, but the North Star just sits there. How is this possible, and why is it so important?

A:

Polaris, the North Star, just happens to sit directly above Earth's axis of rotation. This makes it ideal for navigators, but it won't stay as the North Star for very long ....

One of the really useful things about the stars is that, compared to everything else in the sky, they sit in very fixed patterns. That's because they're so far away.

Indeed, for many thousands of years humans thought of the stars as being fixed on the inside of a bowl or sphere that rotated around the Earth.

Patterns or constellations in the sky are used by navigators—especially at sea—to determine a ship's precise position. And thanks to what is little more than a cosmic coincidence, there is one star that sits right over the North Pole, almost exactly at the point around which the Earth rotates.

It's called Polaris, the North Star, the Pole Star, or the Lodestar. Navigators—at least, navigators in the Northern Hemisphere—have used it for at least a thousand years as a single fixed point from which to take measurements.

Polaris helped us develop more sophisticated navigation and even accurate clocks, because the farther south explorers traveled, the more they could see Polaris shift in the sky. The position of Polaris compared to where it was "supposed" to be let European navigators figure out how far north or south they were. As a result, they then wanted to figure out how far east or west they were—and that's a lot more complicated and requires clocks.

So the coincidence of this star sitting exactly over our North Pole at the exact time we started to become technological enough to build complex clockwork and gearing was really important for the development of our science and theories of the universe!

The funny thing, though, is that Polaris won't stay the North Star for very long. The Earth's orbit has enough so-called "eccentricity" in it, and our rotation wobbles enough that over longer periods—a couple thousand years—the position of the stars in the sky changes quite a bit.

Right now, Polaris will move a tiny bit closer to the exact pole position and then start to move away. By the forty-first century—that's the same amount of time as between now and the Romans—we'll have a new Pole Star called Gamma Cephei.

In fact, Polaris has only been the Pole Star since the twelfth or thirteenth century. It's certainly been close to the pole for the last 1,500 years or more, but the ancient Greeks, for instance, don't talk about it.

A famous Greek navigator named Pytheas, when he described his map of the sky in 320 B.C., said the north celestial pole—where Polaris is now—was empty.

However, Polaris has been a star that never sets (i.e., never dips below the horizon) since at least the fifth century.

We think of the stars as fixed eternal lights in dependable, never-changing patterns (give or take the odd supernova!), but in fact the sky is every bit as fluid as leaves floating on the surface of a pond. Keep watching long enough, and everything changes.

Q:

Why does the Moon always show the same face to the Earth?

Every full Moon looks the same—the same patterns on the surface. Aren't planets and moons supposed to rotate? Why doesn't the Moon?

A:

The Moon does rotate, but at exactly the right speed so that the same face always points to Earth. This is due to the effect of Earth's gravity on the Moon's orbit. But the Moon isn't as still as you might think ....

Let's clear one thing up right away here. You might hear people talk about the "dark side" of the Moon. This doesn't mean that one side of the Moon never sees the Sun, but rather that this side is mysterious because we never see it from Earth. The Apollo missions called it the "far side" of the Moon.

The Moon always shows one face to Earth because it rotates on its own axis at more or less exactly the same speed as it takes to make one orbit of Earth—about 27 days.

Because Earth itself rotates much faster (in one day, obviously!), you can see the Moon moving through its orbit by the way the Sun reflects off it: when the Moon is on the sunward side of the Earth, it's totally black, a so-called new Moon. When the Moon is outside Earth's orbit around the Sun, the Moon is fully lit up in a full Moon. A crescent Moon occurs a little after and a little before a new Moon, with the "horns" of the crescent pointing in opposite directions.

But at all points in the Moon's orbit, the shapes and patterns you can see on the surface are the same. The famous "man in the Moon" and the so-called lunar seas of ancient frozen lava are permanently fixed toward Earth.

This is pretty common for moons around planets. The sheer gravitational power of a large planet on a smaller moon will, over millions of years, synchronize the moon's rotation with its orbit. You can see this "tidal locking" phenomenon all across the Solar System. Most major moons show a single face to their planet—which means if we ever end up living on Jupiter's larger moons like Ganymede or Callisto, only certain colonies will be able to see the beauty of the gas giant in their sky.

When a moon is big enough, it will even tidally lock its planet—Pluto and its moon Charon both show the same face to each other all the time.

Most casual observers think we can only ever see half of the Moon's surface from the Earth, but keener eyes (and astronomers, of course) know different. No moon or planet has a mathematically perfect orbit, and our Moon is no exception. It wobbles and tilts and rocks backward and forward in a process called "libration."

If you take a picture of the moon once every couple of hours for a whole month, then stitch them together into an animation, you won't see a smooth, unmoving disc. You'll see a ball rocking and twisting back and forth, up and down. Over time, you can actually see as much as 59 percent of the Moon's surface from Earth.

Tidal locking is inevitable for most objects in the Solar System. One day in the distant future the Earth may even become tidally locked to the Sun, showing a single face in a single, endless day. But life wouldn't be able to survive in such conditions.

Earth's gravity has locked the Moon's rotation so that its day (one full turn) is the same as its orbit around the Earth, about 27 days

Are the amazing colors in astronomical photos real?

Astrophotography produces some of the most amazing images of our universe, including beautiful galaxies and nebulae filled with stars and swirling gas in many different colors. So why is the night sky just black and white? Why can't we see all that color?

A:

Astronomical photographs are the result of very long exposure times, or multiple photos layered on top of each other. If we lived near a nebula, its real colors wouldn't be nearly as spectacular. But then again, what is "real color" anyway?

One of the happy side-effects of investing billions of dollars of public money in building massive telescopes is coffee-table books full of some of the most amazing images in the natural world.

From nebulae to galaxies and stranger things in between, the cosmos is full of amazing forms, shapes, and above all color. Delicate pastels, blazing blues, deep reds, iridescent greens, all mixed in fantastically shaped clouds of dust and gas.

We've seen such things as the Horsehead Nebula, which by coincidence really does look very much like a horse's head as seen from Earth. Then there's the Orion Nebula, so huge we mistake it for one of the stars in Orion's Belt. And of course there are millions upon millions of beautiful spiral galaxies to photograph.

But are those photographs real, in the same sense as a photo of, say, a flower in a field is a real depiction of what the flower actually looks like? If we took a ride in a spaceship and parked off the side of the Orion Nebula, would we see the amazing detail in the clouds of gas, the blazing young stars, and all those insane colors?

Sadly, with our mere human eyes, we probably wouldn't see much more than a milky glow with perhaps a tinge of red or green. The problem is that objects in space, apart from stars themselves, are very dim—so dim, in fact, that the color-sensitive cone cells in our retinas can't detect them. Instead, we see these objects using our light-sensitive rod cells. But rods can't detect color.

This is, broadly speaking, why the night sky is in black and white. If you concentrate very hard and have good eyes, you might be able to pick out a reddish tinge to Mars or a sort of butterscotch yellow to Saturn. But almost every star looks white to the naked eye.

If we were in a spaceship much closer to a blue or red star, we'd certainly notice the difference in color. Living on planets that don't have exactly the same kind of sun as ours could be a big challenge for humans, just because of the different tint to the light.

Meanwhile, those beautiful nebulae and gas clouds are still out there, glowing dimly in space. So why do they look so amazing when photographed through a large telescope? Are astrophotographers using artistic license and adding color?

Not at all: the colors in those nebulae are really there, scientifically speaking. If you measure the wavelength of the light, it's definitely true that some of it is green, some blue, some red. It's just not very *much* light.

The simple fix for low light when taking a photograph is to increase the exposure. But that then makes the brighter parts of the nebula *over*exposed, washing out the detail.

Astrophotographers therefore tend to use filters and take multiple pictures. First they'll just photograph the red light, then the green, then the blue. When those photos are layered together, then the amazing detail is revealed.

That's not fake detail—the dust and gas really are in there, making those shapes. It's just that our eyes, evolved under a bright yellow sun, aren't sensitive enough to see it without help from our amazing technology.

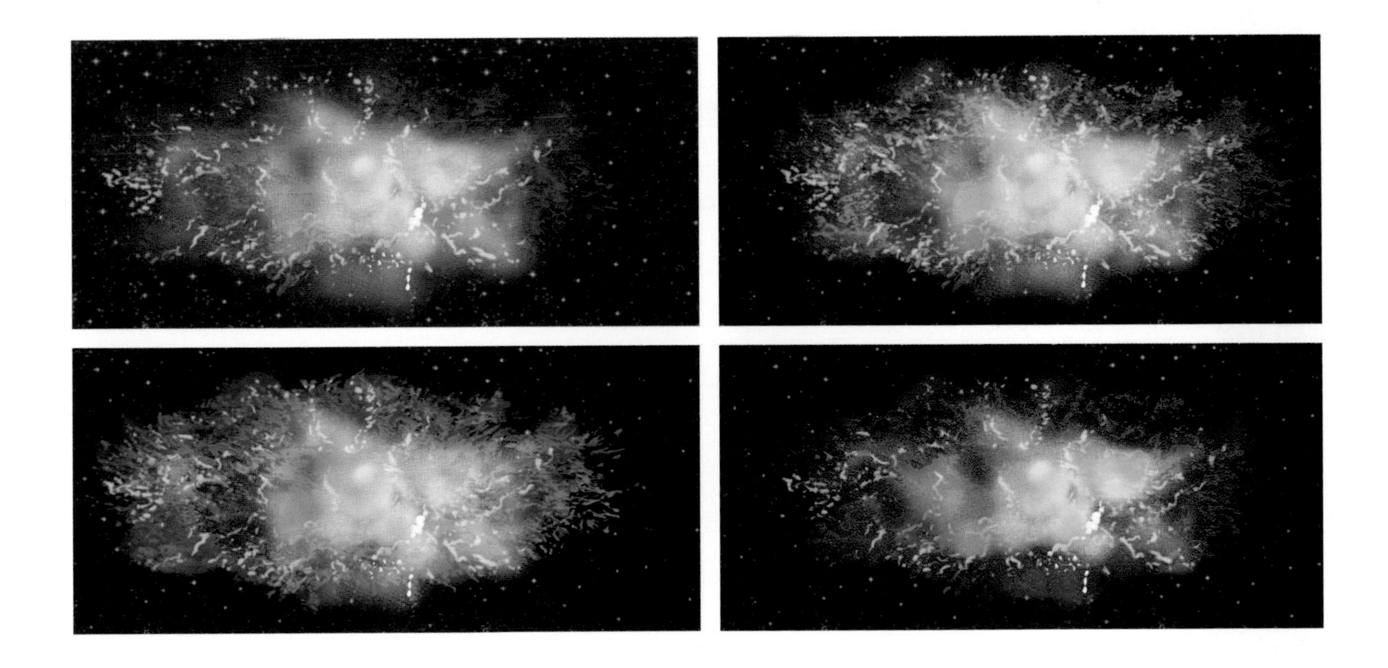

Q:

Is the universe really infinite?

Infinity—numbers without end—is a useful math concept, but can it actually exist in reality? Everything has to have some kind of end, so why do we say the universe is infinite?

A:

Mostly we say it's infinite for convenience. Many equations and problems in physics have simpler answers in an infinite universe. But the actual size of the real universe is much less clear ....

It can seem like science exists mainly just to hang numbers on things. We have all sorts of numbers for all kinds of concepts: electrical charge, gravitational pull, mass, and more. We measure obsessively, giving values to everything from the amount of water in the average human (about 60 percent), to the weight of our entire planet (1.32 times 10 to the power of 25 pounds—in short, a lot).

But there's one number that's really hard to pin down, and that's a value for the size of the universe.

Part of the problem is that we can't "see" the whole universe. Cosmologists call what we can see the "observable universe." The universe is 13.8 billion years old, so you might think that the most distant thing we can see is 13.8 billion light years away. Time itself hasn't existed long enough for the light from more distant objects to reach us.

But because the universe is expanding, we can see farther than 13.8 billion light years. According to cosmologists, the expansion of space puts the most distant object (a source of radiation called the Cosmic Microwave Background) at about 45.7 billion light years away.

That's a big distance, but it's not infinity. There's more universe behind that 45.7 billion light year barrier, and if we traveled, say, a billion light years in any direction, we'd be able to see it. As far as we know, no matter how far you travel, you're always in the middle of a sphere of space 93 billion light years across.

This sort of implies that the universe is infinite. But cosmologists are used to thinking in higher dimensions, and when you cut them they bleed pure math. So to a cosmologist the answer to "is the universe infinite?" isn't straightforward.

Once, humans thought they lived on an infinite flat surface. The world was vast, and early humans assumed it went on forever. Okay—some people thought the world ended in a great waterfall or fire or monsters, but most people just assumed the world went on forever.

Early scientists, curious about this, made observations of how the patterns of the stars changed the farther you traveled, and eventually realized the Earth was a finite globe. Keep walking (and swimming) in one direction and you'll eventually end up back where you started.

This *could* be how the universe works. You might be able to jump in a spaceship, travel for billions upon billions of light years, and end up where you started. But if it really is what cosmologists call a "closed" universe, it's so huge that even when we do experiments on the shape of space, it looks infinite.

Our current level of scientific knowledge isn't advanced enough to definitively answer whether the universe is infinite or not. The true shape of space and time could be much stranger than we think, though at the moment experiments suggest that it's pretty flat. But then, when you stand on the beach and look out to sea, the world looks flat. It's only when you spot the mast of a ship coming over the horizon that you realize you're standing on the surface of an enormous sphere.

Something like that—only much weirder—could be true of the universe.

The Universe

Q:

Is there any actual evidence the Big Bang really happened?

The idea of the whole universe exploding out of a single, infinitely small point sounds pretty far-fetched. How did scientists come up with this wild idea in the first place?

A:

Rather than come up with an idea and go looking for evidence to support it, scientists actually developed the Big Bang theory to explain evidence they already had. And there is a *lot* of evidence ....

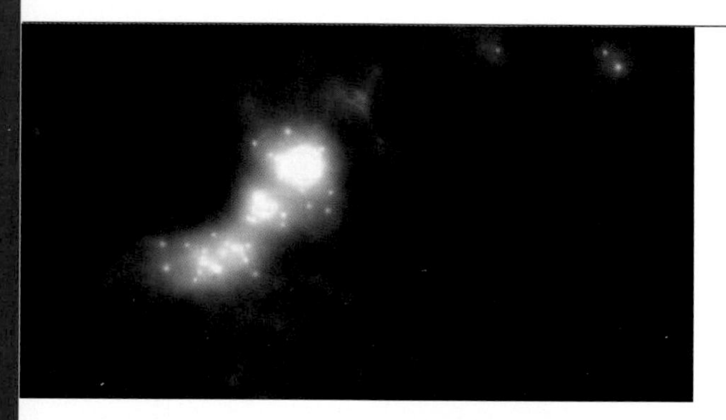

Sometime in the last 400 years or so, humans invented telescopes that allowed us to look into space with incredible detail. And our knowledge of how the universe is built is getting more sophisticated, almost by the day.

One big puzzle was to figure out how the universe began, and how big it was. For a long time, it looked as if the universe had simply always been here. This infinite space, full of stars, had no beginning and no end.

But in the middle of the twentieth century, astronomers discovered something remarkable: objects in deep space were moving away from us, and away from each other. Lots of experiments and confirmation later, and scientists eventually realized that, on average, almost every galaxy in the universe is moving away from every other galaxy. Since galaxies are moving apart, it's simple enough to imagine that at some time in the distant past they must have been closer together. Much, much closer together.

By looking at the way galaxies are moving, and especially because there appears to be no central point from where all the galaxies came, we came up with a remarkable theory.

At some point, every single piece of matter in the universe must have been squashed together into an unthinkably tiny point. And not just the matter. The very dimensions of space itself—height, width, breadth, and even time—were all smooshed together.

Don't worry if this doesn't make a lot of sense. There's a whole bunch of extremely complex math that backs up this theory, and the whole idea of the Big Bang is being constantly refined. The invention of powerful supercomputers has allowed cosmologists to run all sorts of simulations to see how the Big Bang could have really worked.

Part of the problem is the name: Big Bang. It implies a huge explosion, but it was more like someone blowing up a balloon really fast. And the universe isn't *inside* the balloon, it's the *surface* of the balloon.

If you take an empty balloon and draw two dots close together on it, then blow up the balloon, you'll see the dots move apart from each other. That's sort of how the universe expanded—except with more dimensions.

Great theory, right? Galaxies are moving apart, so they must have once been together. But is there any *other* evidence? A smoking gun of some kind? Yes!

Big Bang theory says that in the first few moments after the expansion started, the universe should have been incredibly hot. To us, it would have looked pure white, a blinding light from radiation that would have disintegrated our very atoms in an instant.

So there should be some leftover evidence from this time, a dull glow visible from every point in the universe, no matter where you're standing. And there is. You need sensitive instruments to pick it up, but scientists thought up the idea in their models, went looking for it, and found it in 1965. It's called the "Cosmic Microwave Background," and it's the best evidence we have that the Big Bang, or something very like it, really happened.

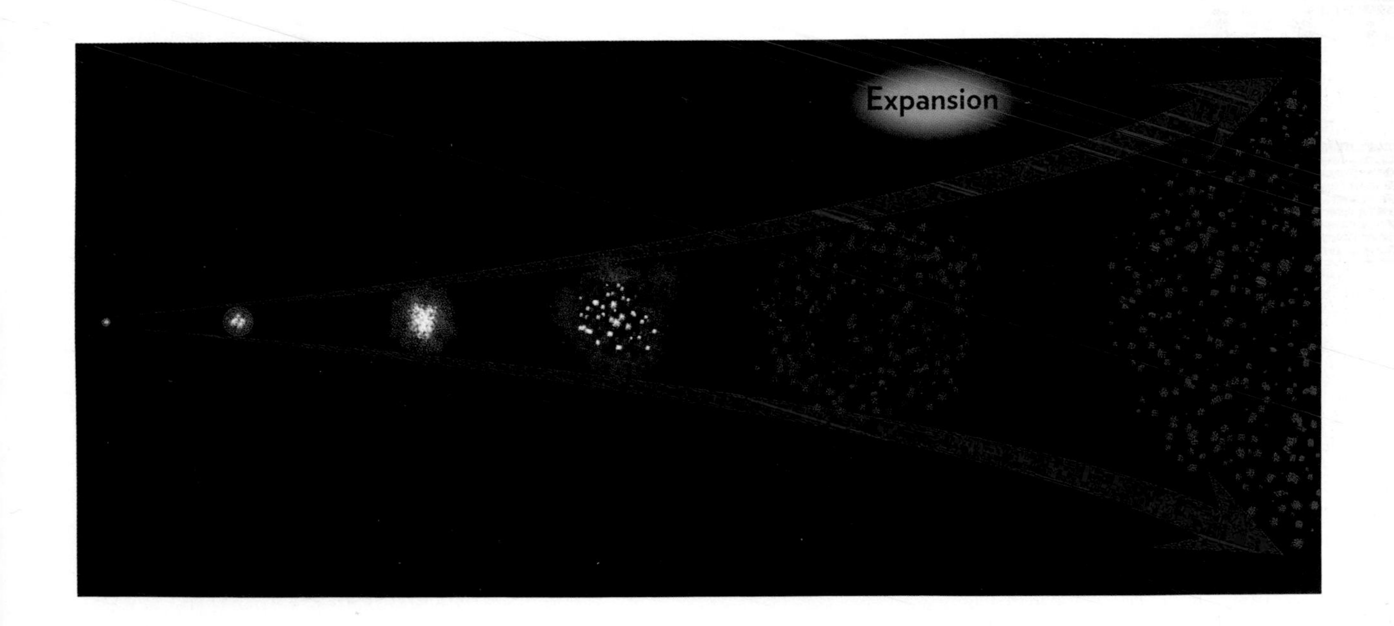

Cosmology

Q:

When and how will the Sun die?

The Sun is basically a giant thermonuclear bomb exploding in space, so eventually it must run out of stuff to explode. When will that happen, and what will it be like?

A:

The good news: our Sun is too small to go supernova and blow up the Earth. The bad news: it may expand and swallow Earth anyway. But it won't happen anytime soon ....

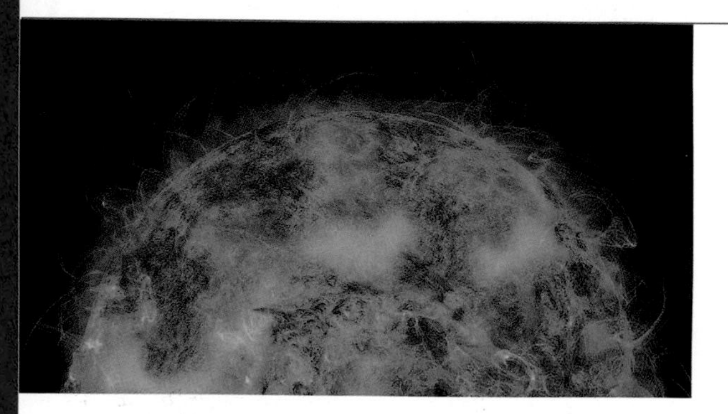

Most of us are taught in school that our Sun is nothing special. An average-sized, averagely bright yellow star somewhere in the middle of the galaxy.

Recent observations have upgraded our humble star, though. It turns out the Sun is brighter than around 85 percent of the stars in the Milky Way. And of the 50 stars closest to us, the Sun is the fourth most massive. Go, Sun!

It's true that the Sun is one giant thermonuclear explosion. Energy from this energy radiates out in every direction, and only a tiny fraction of it actually hits Earth. If we could capture the entire energy output of the Sun for one single second, it would power our civilization for around five million years.

The reason the explosion doesn't just expand out into space and dissipate is because gravity holds the Sun's physical stuff together. Gravity contains the explosion inside a sphere 864,327 miles across. There's so much matter in the Sun that the force of gravity starts the nuclear explosion in the first place by squeezing hydrogen atoms until they fuse into helium and, in doing so, release energy.

In a way, the Sun is slowly eating itself alive. Every time two hydrogen atoms fuse into a helium atom, that's a little less fuel for the Sun to use. Over a very long period—around another 5.4 billion years, according to our current models—the Sun will have used up so much of its hydrogen that extremely dramatic changes will start to occur.

Some very heavy stars, at this point, become so unstable that their internal nuclear fusion reactions are big enough to overcome gravity. Then the whole thing explodes in a spectacular fashion made famous by all good sci-fi disaster novels: a supernova!

Fortunately for us, the Sun isn't big enough to "go nova." Instead of exploding dramatically, the Sun will expand, over many millions of years, into a vast sphere called a red giant. Even though at first this red giant will weigh about the same as the Sun does today, it will be much less dense. And it will be big, really big. The width of this star will be almost exactly the same diameter as Earth's orbit. Our planet might even get swallowed—Mercury and Venus almost certainly will. For more on the Earth's fate, flip to the next page.

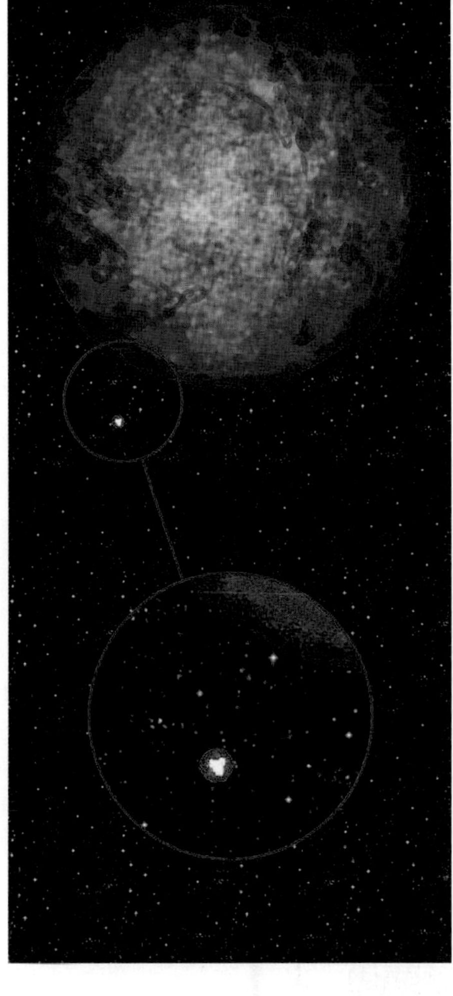

The Sun as a Red Giant (diameter = 2AU = Earth's Orbit)

The Sun as a Main Sequence Star

Over the next billion years, the Sun will lose about a third of its mass while remaining a red giant. Then, after a spectacular ignition called a "helium flash," the Sun will contract back to about 10 times its current size, but will burn 50 times as bright. Because it will have a redder color than today, astronomers call this the "red clump" phase.

After a few more million years, the Sun will expand again, burn very brightly, contract, and shrug off a cloud of matter into a beautiful ring of gas called a planetary nebula.

Finally, our once bright yellow Sun that is the source of all life will contract into a so-called white dwarf. It's almost exactly like a cinder—it doesn't make new heat through nuclear fusion, it just slowly radiates stored heat. But the white dwarf will last for trillions of years, possibly all the way to the end of the universe.

Q:

What will happen to the Solar System (and Earth) after the Sun dies?

Over the next 5.4 billion years, the Sun will slowly expand into a red giant. Will this destroy all the planets, or just the smaller, rocky worlds?

A:

Our Sun is too small to explode in a supernova, but Mercury and Venus will one day be consumed. Earth's fate is less certain, but life here will end long before the Sun dies ....

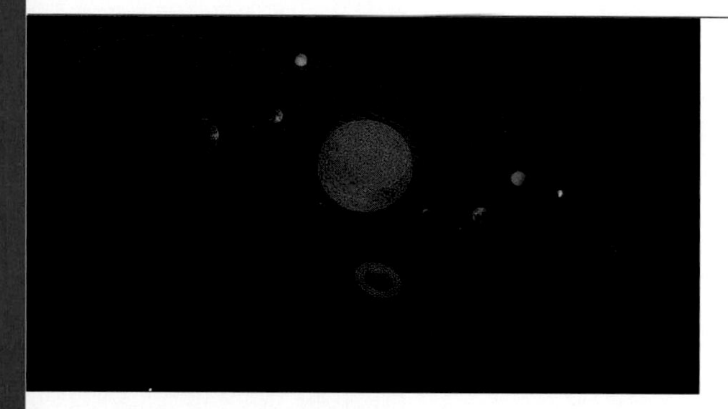

It might seem like the same Sun coming up each morning. But in fact, the new Sun you see each day is a tiny bit brighter than the Sun of the day before. Our star is increasing in brightness as it burns its nuclear fuel of hydrogen. Every billion years or so, the Sun's luminosity increases by around 10 percent.

That means in the time of the dinosaurs, which began about 230 million years ago, the Sun was more than 2 percent dimmer than it is now.

The Earth is a warm world, but even so, the *coldness* of Earth's atmosphere is hugely important to life. Because the temperature of our atmosphere drops below freezing only a few thousand feet from the surface, water vapor (water in gas form) condenses into liquid and rains back into the ocean. Our water is essentially trapped.

As the Sun gets brighter and thus hotter, Earth's upper atmosphere will warm, and water won't condense as quickly. We'll start to lose water to space, just boiling off into the void. After about a billion years, our oceans will dry out. Without liquid water on the surface, life can't survive, no matter the temperature.

But maybe human technology will be so advanced that we'll be able to stop this happening. Maybe we'll keep our ecosystem intact long enough to watch the Sun expand into a red giant. What then? Will the planet just get swallowed up?

Maybe not. By the time the Sun expands into its giant phase, it will have converted nearly 30 percent of its mass into energy. Since it weighs less, the Sun will have less of a gravitational pull on Earth, and our orbit will move farther out—nearly twice as far as we are now.

Unfortunately, that might not be enough to save the planet. For a start, the Sun will be incredibly bright—many hundreds and eventually many *thousands* of times brighter than it is now.

What's more, tidal effects similar to the way the Earth and Moon interact today could slow down our orbit. The Earth will fall back toward the Sun, enter the upper layers of the red giant, and be slowly vaporized.

That will be that for Earth in the year 5 billion A.D. (give or take a few million years). The gas giant planets including Jupiter and Saturn will survive, and their moons—frozen today—will be bathed in the light from a much brighter Sun, and become much warmer. On these pocket-sized worlds, with new shallow liquid oceans, life might find a new home.

But if life does find a place on, say, Jupiter's moons of Europa, Callisto, and Ganymede, or Saturn's great moon Titan, it won't be able to live there forever.

The Sun will only stay a red giant for around a billion years, and then it will begin a relatively fast process of collapsing into a white dwarf. On the way it will cast off a planetary nebula, a ring of gas that could knock planets off their orbits and send them wandering the galaxy.

But one day, the gas and dust left behind by our dying Sun could mix with other material in the so-called "interstellar medium," clump together, and start the process all over again by forming a new star.

Indeed, that's exactly how our Solar System formed in the first place.

Q:

Will the universe ever end?

If we accept that the universe began with the Big Bang, does that imply that it must one day end? How far off is that time, and what will it be like?

A:

We don't know for sure how the universe will end, but we have some pretty outlandish theories. Everything might just grind to a halt, or the very fabric of space and time could pop like a balloon ....

For most of the eighteenth and nineteenth centuries we thought the universe was eternal: had always been and always would be.

Then we realized space itself is expanding, which implies everything used to be crammed into one tiny spot, which in turn implies the universe had a beginning. We call that beginning the Big Bang, and if the universe had a beginning, then it could also have an end.

For most of the twentieth century, we thought that the force of gravity would eventually overcome the force that causes the universe to expand. Like they were attached to some kind of celestial elastic band, galaxies would slow, then fall back toward each other, eventually all cramming together again in an event called the Big Crunch. That would be the end of the universe.

Unfortunately for people who like neat answers, in 1998 we discovered the expansion of the universe is actually speeding up. If our math is right, gravity will never overcome the force of expansion, and the universe will just get bigger and bigger until ... well, we're not sure.

One possibility is that the rate of expansion will get so high, something called the "Big Rip" will happen. It's almost like blowing up a balloon until it pops—except with a lot more dimensions and quantum physics.

It's also possible that the whole universe will just run down. Stars will use up all their nuclear fuel, black holes will evaporate, and the whole cosmos will settle down into a perfectly even temperature.

If there are no places in the universe with concentrations of energy, that energy can't be used to do work or computation or to support life. There will just be nothing ... a slightly tepid nothing. This idea—called the "Heat Death of the Universe" comes from the physical laws that scientists use to describe the way energy moves around a physical system.

When you ride a bike, chemical energy flows out of your cells into your muscles, which move and put that energy into the bicycle. The wheels push against the ground, and some of that energy is converted into heat (from friction) and some into motion. Eventually your cells run out of the kind of energy they can use to move your muscles, and you have to rest or eat something. But if you have no way to get more energy, you can't keep riding your bike (also you will ultimately die, but let's not get dramatic).

The problem with this explanation is that while it fits very well for "small" things like people, airliners, earthquakes, planets, stars, and even galaxies, it might not fit a system as big as the whole universe.

Today, there are lots of different theories about the universe's ultimate fate. Maybe we're living on the skin of a previous universe. Maybe gravity will stop everything running down. Then there are concepts like "dark matter" and "dark energy" that complicate our model even more.

The irony is that for thousands of years, when humans didn't know much about science, we were very confident that the universe would never end, or that it would end in some kind of apocalypse. Then, as we developed science, we were confident the universe was eternal. Then, as our science got better we became ... less confident ... until today, it seems like the more we learn, the less certain we can be about the ultimate fate of our universe.

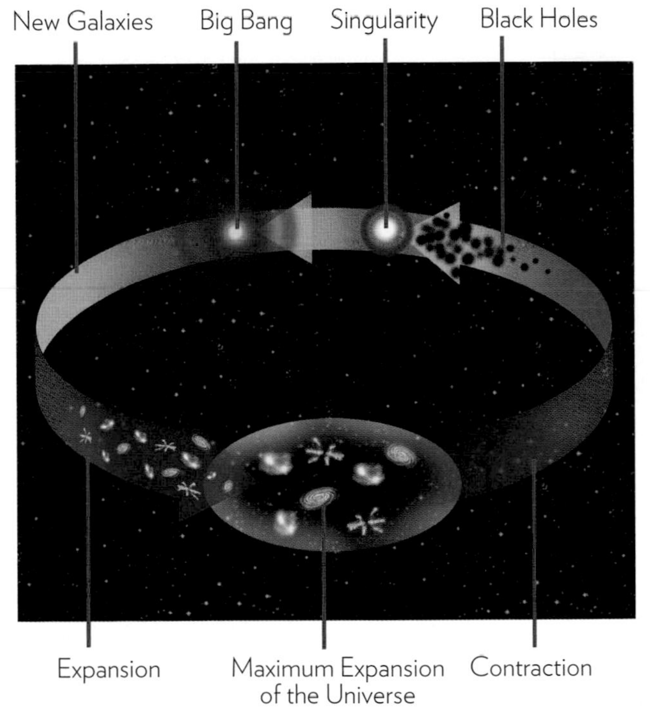

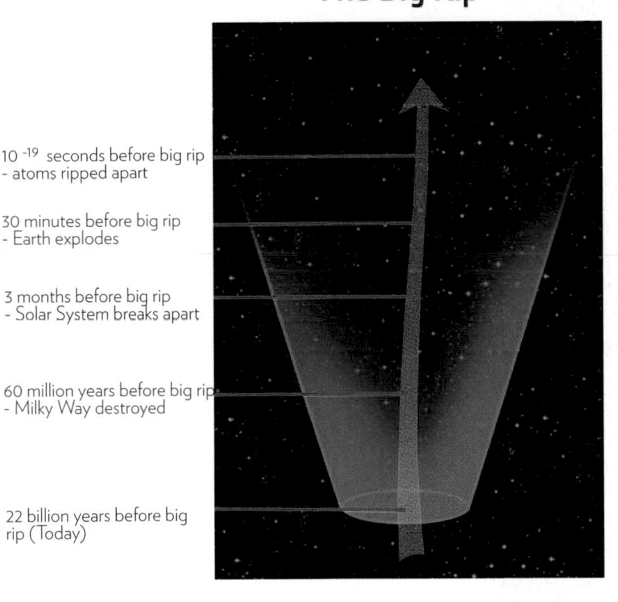

Q:

If there really were aliens on other planets, wouldn't we have met them by now?

We're discovering more and more Earth-sized planets, but we still haven't found any evidence of other intelligent life. Where are the aliens? Are we really alone in the galaxy?

A:

While it's puzzling that we haven't seen any aliens in our stellar neighborhood, we've only surveyed a tiny fraction of the galaxy. It's just too early to tell ....

The case of the missing aliens has a name: the Fermi Paradox. An Italian physicist named Enrico Fermi came up with it in 1950. He said, since the galaxy is so big and so old, there must be other Earth-like planets that could have evolved life millions of years before us. Even if faster-than-light travel is impossible, over millions of years a civilization could easily colonize the whole galaxy. So where's the evidence of that? One of the problems with answering this question is that humans are pretty bad judges of scale. We think a century is more than a lifetime. We take one quick glance at the sky and say: "Nope, can't see any aliens, must be no aliens."

The Fermi Paradox assumes that since we can't find evidence of aliens really easily, there must be no aliens. It doesn't take into account the fact we really haven't been looking for very long, or very far.

You might think that because we've shot probes past all of the planets, we've pretty much explored the whole Solar System. But we've only seen a tiny fraction of what's in our own backyard. There could be alien probes observing Earth right now from, say, the Asteroid Belt, and we wouldn't be able to tell.

Many people also think we should be able to pick up radio signals from other civilizations around other stars, that the sky should be full of radio chatter and alien TV stations and suchlike.

They think this because Earth has indeed been broadcasting radio into space for nearly 100 years. One of our earliest broadcasts was Adolf Hitler opening the Berlin Olympic Games in 1936. Science-fiction writers have had a lot of fun with the idea of aliens picking up that signal and sending it back to us ....

But these signals aren't that powerful when you're talking interstellar distances. Earth's own signals become almost invisibly weak only a light year from the planet, and the nearest star—Proxima Centauri—is four light years away. Picking up the signal over the general background radio noise of the universe when you don't even know it's there is pretty much impossible.

There could be a civilization on a planet orbiting Alpha Centauri right now, with satellite TV, and our current technology wouldn't be able to detect it.

What Fermi's Paradox is really asking is why there don't appear to be any aliens who've built giant hyperspace webs or ringworlds or other exotic sci-fi constructions that are visible at interstellar distances. Things we can see without even trying. There are lots of possible explanations for this, from galaxy-wide natural disasters to humans just being too stupid to recognize an artificial star even when we're staring right at one.

Is it possible we are the only planet in the entire universe that has evolved intelligent life? The odds seem stacked against that idea. Current estimations suggest there are likely to be more planets in our galaxy than stars—over 300 billion. Only a fraction of them will be Earthlike, but that's still millions of planets. To assume all of them are uninhabited after barely 100 years of serious observation just seems silly.

Q:

Why isn't Pluto considered a planet anymore?

Poor Pluto. It used to get on all the posters of the Solar System as the ninth planet. But then in 2006, it was demoted to a mere "dwarf planet" Was this fair? Why does it even matter?

A:

There's no good single reason to *not* call Pluto a planet. But the problem is, there's no good single reason *to* call it a planet either. It's not even the biggest dwarf plant in the Solar System ....

One of the most surprising public reactions to a scientific announcement came in 2006 when the International Astronomical Union first hinted at and then went ahead with changing Pluto's classification from "planet" to "dwarf planet."

People were really angry about this, probably because they'd gone through school being taught there were nine planets in our Solar System. Suddenly, we were back to eight.

Actually, for most of human history there were only five planets visible from Earth—Mercury, Venus, Mars, Jupiter, and Saturn. Uranus is visible, but it's so faint you need to know where to look—we didn't confirm its existence until 1781.

Neptune was only discovered in 1846, not by observation, but by math. Astronomers, puzzled by Uranus' weird orbit, theorized the existence of an eighth planet. Once they figured out Neptune's orbit, they went looking and sure enough there it was.

In more or less the same way, astronomers predicted the existence of Pluto. After much careful and painstaking checking of astronomical photographs—it's been said the search was like looking for one particular grain of sand on a beach—Pluto turned up. The ninth planet had been found!

Unfortunately, as the twentieth century went on and we launched orbital telescopes like Hubble, we began to discover even more small planet-like objects in the far reaches of the Solar System. One of them, called Eris, is even slightly bigger than Pluto. This makes things confusing.

What if there are dozens of Pluto-sized objects in the outer Solar System? Can we just call them all planets? Nine planets, okay, but twenty? Thirty? It didn't help that at the time there was no formal definition of what a planet is. So the International Astronomical Union decided that a planet was any object that directly orbited the Sun, was spherical, and—this is the tricky part—had cleared its orbit of all other objects (by smashing into them, capturing them as moons, or ejecting them).

This is the point on which Pluto "fails" to qualify as a planet. There's a huge group of objects all moving around the Sun in Pluto's orbit. Pluto has a surprisingly large number of moons and weighs just 0.07 times as much as everything else that orbits with it. The Earth, on the other hand, is 1.7 *million* times heavier than the few bits of dust and specks of rock left in our orbit.

The thing is, though, this isn't the first time something like this has happened. The largest of the asteroids, Ceres, which orbits between Mars and Jupiter, is a sphere 590 miles (950 km) across. When it was discovered in 1801, it was classified as a planet, and it stayed a planet for about 50 years (Pluto was a planet for 76 years). Then it was reclassified as an asteroid. Now, more than a century later, Ceres has joined Pluto as a dwarf-planet. So Pluto got demoted, but Ceres got promoted! Yay for Ceres!

At the end of the day, these labels are all just for purposes of scientific classification. You're of course free to call Pluto whatever you want. It's a great way to get into an argument with an astronomer.

Q:

Is the Andromeda galaxy really going to crash into the Milky Way?

The evidence is pretty clear: the Andromeda galaxy is moving toward the Milky Way, getting closer each year. Could the two galaxies collide, and what would that be like?

A:

They will collide, but it won't be violent like a car accident. Though it could seriously disrupt life on whatever planet we're living on at the time ....

One of the first galaxies we ever discovered is called the Andromeda galaxy. It lies in the constellation Andromeda, and we used to think it was a nebula until astronomer Edwin Hubble realized it was something much bigger.

Andromeda is a spiral like the Milky Way but has about three or four times as many stars—a trillion at least, maybe many more. Currently, it lies about 2.5 million light years away.

Observations using the Hubble Space Telescope have led astronomers to a near-certain conclusion that Andromeda and the Milky Way will collide and merge in about 3.75 billion years.

But don't expect a huge explosion and a lot of smashed-up planets. The distances between individual stars and worlds in these galaxies is huge.

Even if the Sun were as big as a golf ball, the nearest star, Proxima Centauri, would still be 680 miles (1,094km) away.

So really we should say the galaxies will merge, rather than collide. With trillions of stars in the mix, there is a good chance there will be some catastrophic results—not from actual collisions, but from the way gravity will disrupt the stars' movement. Many will be hurled off into space, to begin a long, slow journey back toward the new supergalaxy.

Our own Sun could be dragged much closer to the galactic center, or pushed much farther out. If the collision happened today it probably wouldn't make much difference to conditions here on Earth. When it does occur, 3 to 4 billion years from now, it's unlikely our planet will still be inhabitable anyway, because the Sun will become much brighter by itself and will boil off our oceans.

The other thing to remember is this process will happen very, very slowly, over millions of years. Whole civilizations like ours could evolve and die out on worlds without even really noticing the galaxies are colliding.

It's not even unusual: there are several dwarf galaxies currently merging with the Milky Way, and the shape of our galactic center suggests we've already merged with at least one reasonably large galaxy at some time in the distant past. We have lots of pictures of other large galaxies elsewhere in the universe merging right now.

Scientists use supercomputers to imagine how Andromeda and the Milky Way will merge, and they even have a name for the new galaxy: Milkomeda.

Right in the middle, the two galactic centers are likely to join together and form an incredibly intense ball of radiation around a supermassive black hole. This is called a quasar, and it would be as bright as the full Moon, seen from Earth.

There's one other large, spiral galaxy in our local group of galaxies. It's called Triangulum, and it would be affected by the merger of the Milky Way and Andromeda, too. It will end up orbiting the new galaxy, having stars stolen by its new, larger sibling.

The worst thing about the Andromeda/Milky Way collision is that the beautiful spiral arms of our two galaxies will be blurred and erased. Instead, we'll live in an elliptical galaxy, a vast disk of trillions of stars.

Q:

How many Earth-like planets could there be?

Scientists discover new so-called exoplanets every day, it seems, but many are described as "super Jupiters." Are there any Earth-like worlds out there? How many?

A:

There could be millions, if not billions. The latest results of planet-hunting missions suggest there may be as many planets in our galaxy as there are stars. Over 100 billion ....

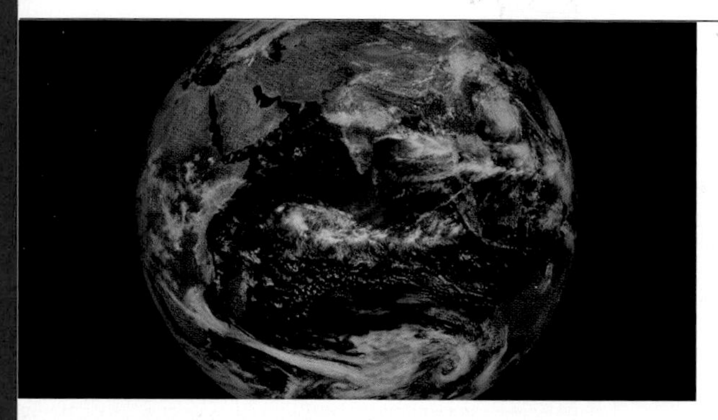

It's strange to think that despite all the hundreds of sci-fi novels and movies set in other worlds, we actually don't have any undeniable scientific proof of the existence of a second Earth.

By second Earth, we mean a dense rocky world with an oxygen-rich atmosphere, liquid oceans, and a temperature more or less compatible with the kind of life we find here at home.

Once we developed the technique to find planets around other stars and started hunting for them in earnest with instruments like the Kepler satellite, we discovered that having planets is very much the norm for the average star in the Milky Way galaxy.

We've already found hundreds of worlds, but most of them are as far from Earth-like as you can imagine. And some even farther. How about the world where it rains molten glass, for starters?

But this constant discovery of planets almost everywhere we point a telescope makes it seem very probable that most stars will have a collection of worlds orbiting them, and that the planets close in will be small and rocky.

So yes, as this book reaches you there haven't been any absolute, no-question Earth-like worlds discovered yet. We have found some so-called "super Earths," worlds that are much larger than ours but that seem to have the same mass. They weigh as much as Earth, and it's likely that they will have solid surfaces like ours. Will they have oceans, continents, and life? It's far too early to say.

But as planet-hunting astronomers continue their search for exoplanets (an exoplanet is any world not in our home Solar System), they can start to make assumptions based on what they've already found. By using probability mathematics, scientists can make very educated guesses about how many planets there are in the galaxy.

And the answer is that there could well be many more planets than there are stars. The Milky Way has around 100 billion stars, so that's a lot of worlds.

Our Solar System alone has three worlds that, strictly speaking, are Earth-like. Sure, Venus has a crazy runaway greenhouse-effect atmosphere, but the planet itself is almost exactly the same size and density as Earth. Mars is quite a bit smaller (38 percent of our surface area, 10 percent of our mass), but shows evidence it once had liquid water on the surface. Even with today's science, we can imagine geo-engineering Venus and Mars to make them fit for human habitation.

We can look out into the Milky Way and see stars that are very much like the Sun. And we can see stars with planets orbiting them. It's pretty basic logic that if planets are common, then a star like the Sun is likely to have planets similar in composition to our Solar System.

And this is just in our galaxy. There are hundreds of billions of galaxies in the part of the universe we can see. So the true number of Earth-like worlds could be, without exaggeration, virtually uncountable.

= Millions or billions in the Milky Way galaxy alone

Q:

Why are pulsars so important to astronomers?

Pulsars are rapidly rotating neutron stars that pulse out intense beams of radiation at precise intervals. That sounds cool, but what's the real significance of these weird objects?

A:

Because of their incredibly regular rotation, scientists can use pulsars for everything from keeping super-accurate time to mapping the galaxy.

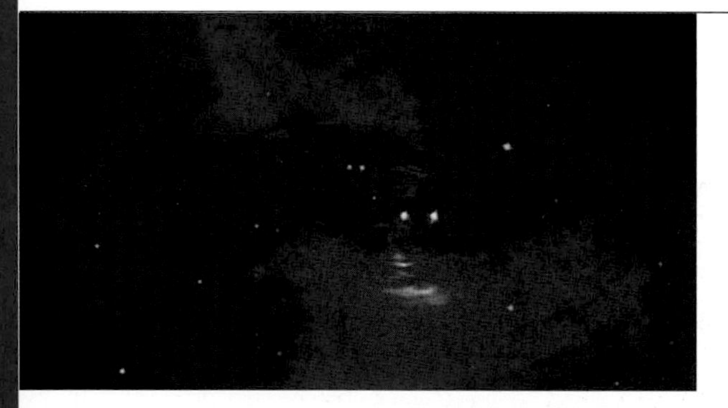

On November 28, 1967, astronomers Jocelyn Bell Burnell and Antony Hewish made a remarkable discovery. In the constellation of Vulpecula, they observed what they could only describe as a radio beacon, blinking on and off every 1.33 seconds.

It was so precise and so regular it baffled the astronomers. They even thought, just for a moment, that it might be a signal from an extraterrestrial civilization. So they gave it the designation LGM-1, which stood for "Little Green Men."

When the astronomers announced their discovery, other scientists quickly realized that LGM-1 was in fact a rapidly spinning neutron star. A natural phenomenon, easily explained without the need for aliens—little green ones or otherwise.

When stars die, they can end up as several different kinds of object. The most massive stars collapse into black holes. Smaller stars like our Sun end up as slowly cooling white dwarfs.

But stars of just the right mass—between 1.4 and 3.2 times as massive as our Sun—collapse into a strange sort of superdense matter called a neutron star.

These stars are only 7 to 8 miles (11 to 13km) across, but they weigh 500,000 times more than Earth. To get an idea of how dense a neutron star is, imagine a fairly big luxury cruise liner crushed down into the size of a peanut.

Weird stuff happens to matter when it's so tightly compressed. Neutron stars generate powerful magnetic fields, and they spew out powerful beams of radiation—mostly radio waves, but also X-rays, gamma rays, and even visible light.

Neutron stars also spin. The speed of their rotation depends on how big the original star was. And if the beams of radiation shoot out at an angle from the neutron star's rotation, then the star acts like a cosmic lighthouse.

On Earth, we see the beam of radiation as the neutron star spins. Our instruments pick it up as a pulse, so we call these objects pulsating stars—or pulsars.

Not all neutron stars are pulsars, but so far we've identified about 1,800 in our galaxy. They're proving incredibly useful for astronomers.

For a start, the precise regularity of a pulsar's rotation rivals the accuracy of our best atomic clocks. Scientists can time a pulsar's spin and, well, set their watches by it.

Because each pulsar has its own unique pulse speed, we can use them as points of reference when building maps of the galaxy. When we sent out the Pioneer and Voyager probes, for instance, NASA included a "map" of the location of Earth, based on our position relative to 14 pulsars in our local area.

Pulsars could even end up helping us figure out the final mysteries surrounding how gravity works. When other stars orbit pulsars, cosmologists expect to see evidence of so far theoretical "gravitational waves"—ripples in the fabric of space-time.

If we can get some solid data on gravitational waves, it could help confirm many aspects of our model of the universe that, at the moment, are just theoretical. We've already seen indirect evidence of gravitational waves from a pulsar, and experiments continue.

Finally, one day in the future when human starships are out exploring the galaxy, pulsars could be used for a sort of interstellar GPS, giving the ship a precise location and pointing the way home.

White Dwarf

Pulsar

Black Hole

physics

The universe runs according to a set of rules, and physics is the rulebook ....

Chemistry might give us the "how" for many mysteries of science and nature, but only physics provides the "why." With a set of equations—and a healthy dose of genius—physicists can explain and describe almost everything we see and experience.

Why is light the fastest thing in the universe? Physics knows. How can I prove the Earth orbits the Sun? Physics tells you. Why did I get electrocuted when I stuck a fork in the toaster? Physics will explain it to your next of kin.

From simple fundamental rules comes the amazing complexity of nature. From an equation like "force equals mass times acceleration" comes the fury of a hurricane, or the excitement of a baseball game.

Physics gives us the link between science fiction and science fact, and may one day answer the ultimate question: why do we even exist?

Q: Why can't a spaceship travel faster than light?

Physicists tell us there's a universal speed limit that's impossible to break: the speed of light. Why can't we break it?

A: Not only can we not break the speed of light, we can't even reach it. Einstein's theory of relativity makes it impossible ... unless you're prepared to put on a bit of weight.

On the one hand, you might think the speed of light is pretty fast: it's 186,282 miles (299,793km) per hour. That means there's nowhere on Earth you can't send a signal to virtually instantaneously. But once you leave Earth, that speed limit starts to feel a bit more restrictive.

A conversation with a friend on the Moon will have a 2.6 second delay, because it takes light (or radio waves) about 1.3 seconds to reach the lunar surface—and of course you have to speak, wait 1.3 seconds, then your friend speaks, and you wait another 1.3 seconds. Irritating!

Talking to someone on Mars is much worse: depending on the time of year for each planet, the distance between Mars and Earth can be big enough for a 20-minute delay! And Mars is relatively close ....

The nearest star to Earth, Proxima Centauri, would have an *eight year* delay on any communication, since it's four light years away.

And this is for signals made of radio waves traveling at or very near the speed of light. A real spacecraft can't go nearly as fast and will take even longer. But why this universal speed limit?

Albert Einstein figured out, in the early twentieth century, that energy and mass (or weight) have a close-knit relationship. He came up with the world's most famous mathematical equation: $E=mc^2$. Energy equals mass times the square of the speed of light.

How it applies to our spaceship is that every time we accelerate using, say, a cool fusion engine, the overall mass of the spaceship increases. At the kinds of speeds we've achieved with our technology so far, this increase in mass is almost undetectable, mere fractions of an ounce. But as speed creeps up toward the speed of light, mass starts to increase, well, massively!

In short, the faster you go, the heavier you get. The heavier you get, the harder your engine has to push to make you go even faster. Einstein's equations show that when the ship is a tiny fraction off true light speed, it weighs so much that accelerating it that final bit is pretty much impossible. It would take an *infinite* amount of energy to complete the acceleration.

This is pretty frustrating for science-fiction writers and anyone who would like to visit a neighboring star system before they die of old age. But we can take heart from the idea that this light speed limit might only apply to normal physical objects in normal space, being pushed around by normal forces like rockets or ion drives or gravity.

If we let physicists go crazy with theories and math, they come up with ideas like wormholes, which are strange regions that could link two distant points in space. Like a celestial shortcut, a ship could enter one end of the wormhole and come out the other. Would it get ripped to pieces by weird gravity effects or shredded by powerful radiation? Maybe, but at least the chance of a real warp drive is there!

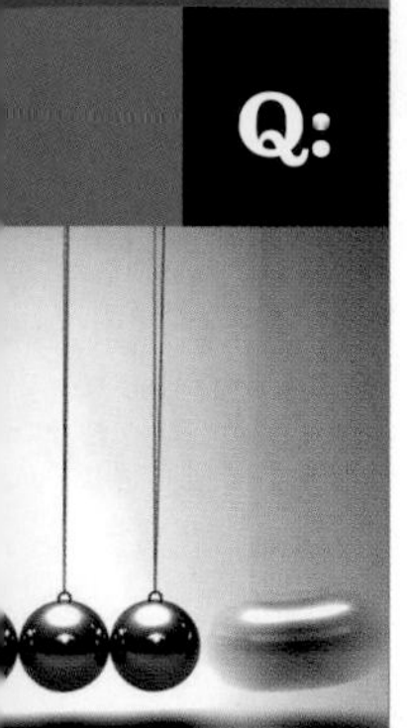

Q:

Is the speed of light the same everywhere?

Everything in the universe can travel at different speeds, or be slowed down and sped up. What about light? Does it always travel at a single speed?

A:

The speed of light in Einstein's famous equation $E=mc^2$ refers to the speed of light in the vacuum of space. Through other materials like air, water, and glass, light does indeed travel at a different speed.

Nothing in the universe can travel faster than light ... based on our current understanding of physics, at least. But that maximum speed is how fast light moves in a vacuum—a completely empty space.

Light can be measured as tiny particles called photons. Like anything, photons are affected by the medium—or substance—they travel through.

The objects we see with our eyes have the shapes and colors they do because of the way photons either bounce off them or get absorbed by them.

Some materials don't absorb or reflect photons at all. Light passes straight through the material, so we can't see it. The Earth's atmosphere is a good example. It's completely invisible to us, but it's still there, and photons have to pass through it.

Our atmosphere is actually thick enough to slow light down very slightly from that universal top speed that physicists just call c. Other materials slow it even more. When light has to pass through an optical fiber such as in a transatlantic telephone call, it's slowed by as much as 35 percent!

We call this phenomenon "refraction" and it allows us to see some objects that would otherwise be invisible. Pure water, for instance, doesn't absorb very many photons at all.

If light passed through pure water at 100 percent of light speed, the liquid would be pretty much invisible to us. But light is slowed down by having to pass through all those hydrogen and oxygen atoms in the water. When light comes out of the water again, it can suddenly speed up in air. This speeding up is visible to us by the light getting bent and changing direction slightly.

You can see this if you put a straight object into the water—it will appear to bend at the surface. This is refraction in action, and it happens because light has to move more slowly through water. Different materials slow light in different ways, and this is measured by a refractive index.

Scientists are fascinated with refraction and have done many experiments to see exactly how much they can slow light down. Some experiments have even managed to stop light altogether, though really what's happening is the light is being stored in the material and will be released later.

The speed of light in different materials—or mediums as physicists call them—has a real effect on our everyday lives. Computers, for instance, shoot electrons around in their central processing chips so fast that the speed of light becomes a limitation.

A modern PC can do more than three billion calculations every second, which is so fast that if it had to send its electrons more than 12 inches (30.5cm), they wouldn't be able to get there fast enough to finish the calculation. Because of this, chips have to be small, and this makes them run very hot. So computers need complex fans and thermal design. All because the speed of light is a little bit slower inside a silicon chip!

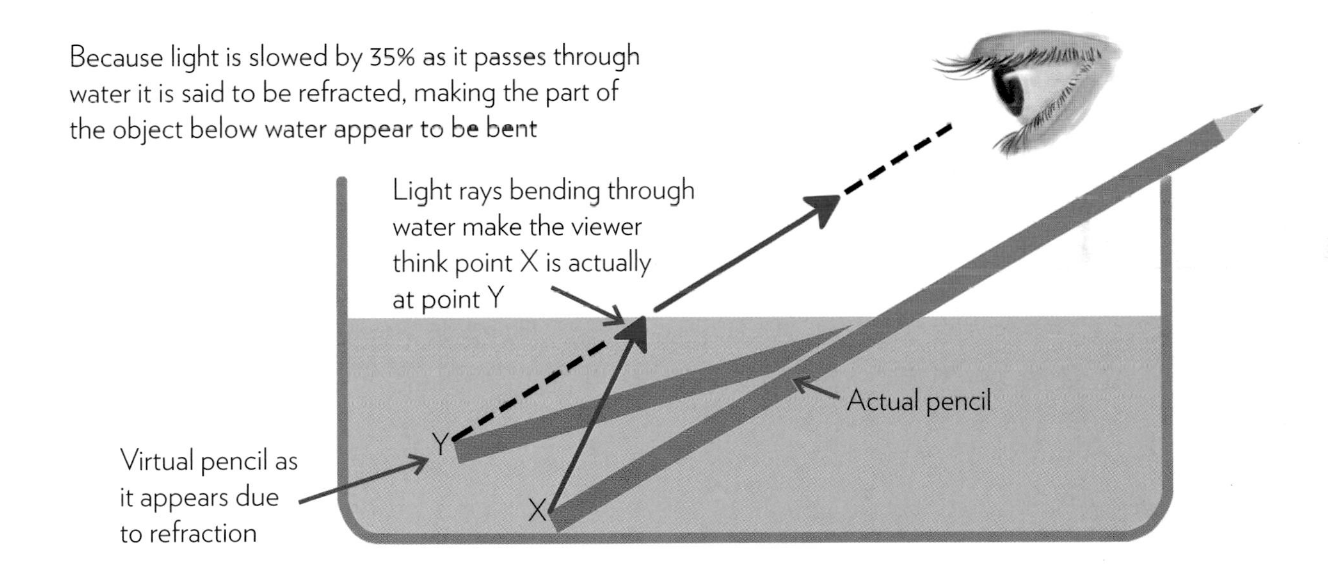

Q:

Why did we invent quantum physics?

Just when you think you've got a handle on how the universe works, along comes quantum physics to make everything so much more complicated. Why do we need all this confusing stuff?

A:

To match their theories with actual observations, scientists had to come up with a new set of natural laws for how matter behaves on a really small scale. The result is ... tricky.

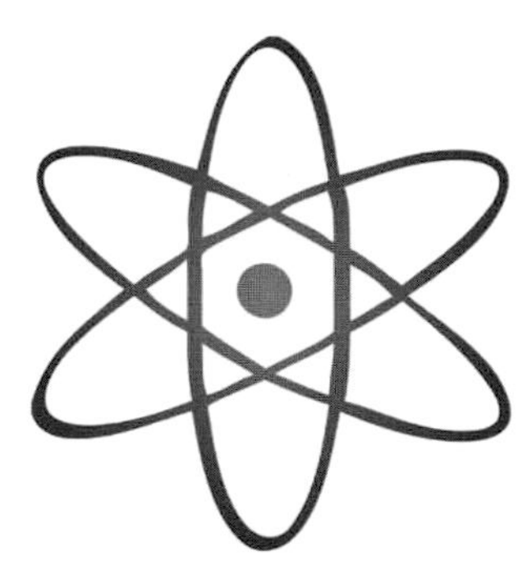

Quantum physics or quantum mechanics began to emerge from a set of theories in the early part of the twentieth century. Famous scientists like Albert Einstein and Max Planck were troubled by some of the observations they'd been making and needed a way to make their theories fit the actual results of experiments.

As our scientific understanding and lab technology improved, allowing us to probe the structure of the universe and discover such things as the nucleus of the atom, protons, neutrons, electrons, and more, it became obvious that the universe is much more complex than we first thought.

The behavior of large chunks of matter—and by large we mean anything from a bacterium to a galaxy—is fairly easy to predict and model with math. Stuff in the universe is affected by gravity, and you can speed it up or slow it down by using or releasing energy. Laws like Newton's Three Laws of Motion and the Laws of Thermodynamics rather neatly explain lots of the big stuff we observe.

But then we started to do experiments with much more sensitive equipment, and we started to try to figure out answers to questions like "What exactly is light?" The results messed everything up.

If you have the energy, you can change the state of large things by arbitrarily different amounts. To explain: if you want to heat up some water, you can heat it up by 10°, or by 7°, or by 7.34°, or by 7.664324°. There's a continuous scale on which you can change things, until you get down to the subatomic level.

Physicists discovered, in the early twentieth century, that at very, very small scales, quantities can only change in certain discrete amounts. They used the Latin word *quanta* to describe this, which is why we call it "quantum physics."

What's more, many fundamental particles such as the photon (which makes light) and the electron (which carries a negative charge and makes electricity work) can be observed as either a distinct particle or as a wave.

In the late nineteenth century, physicists developed wave theory, which worked really well until new experiments showed photons and electrons also acting as particles.

An object in the everyday world has a few essential properties. It has mass (which determines its weight in Earth gravity), it has a size in three dimensions, and it has an absolute position. It also has a speed and a rate of acceleration. All objects have these properties ... unless you look at things on a subatomic scale.

Some particles, like the electron, appear not to have a mass. And others seem to have no distinct position, or can appear in two places at once. These phenomena have been observed in experiments, over and over again, in labs across the world.

To explain all this, the theory of quantum physics was developed. In 1927, it was broadly accepted as being true by the scientific community at large. Today, it's an essential part of our understanding of the universe.

The Holy Grail of physics today is to find a way to unify quantum physics and the so-called "standard model" of the larger universe. If we can figure that out, well, it could literally explain everything.

In a solar system (like ours) the planets can assume an almost infinite number of different configurations since every planet is moving and changes position even after a fraction of a second. This can be explained and described by classical physics. But classical physics doesn't work at the subatomic level.

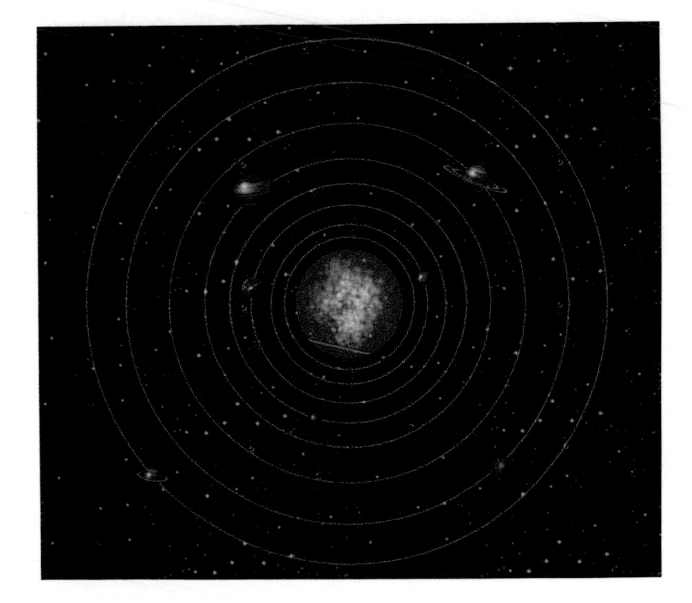

Even though an atom resembles a solar system (this is the atomic diagram for gold) its electrons do not behave like planets orbiting the Sun. They can only be observed in a finite set of different configurations or "quanta." Classical physics can't properly describe how subatomic particles behave, so we need another system—quantum physics.

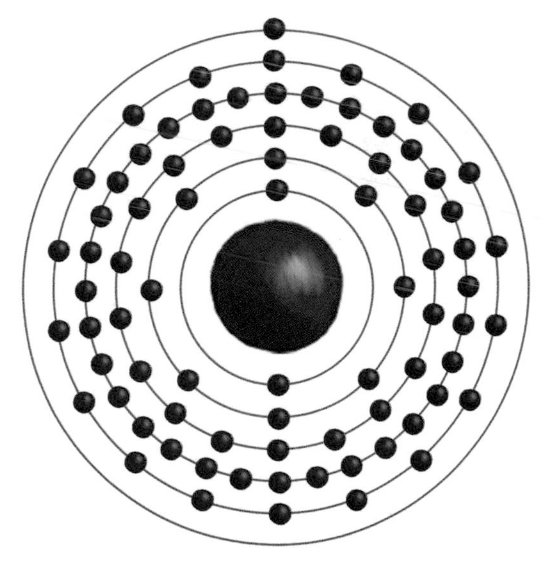

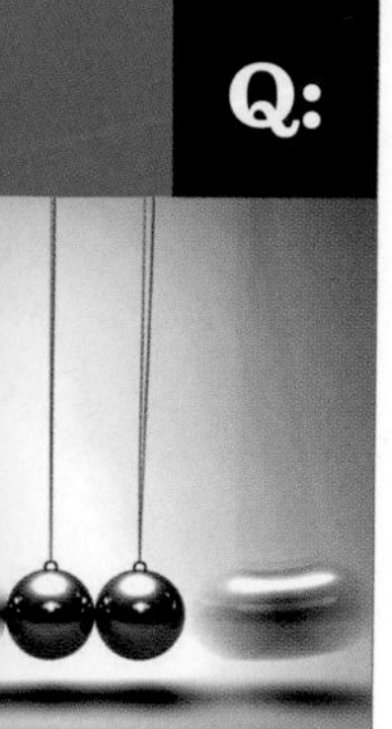

Q:

Is time travel possible?

Having just learned this week's lottery numbers, I'd like to travel back in time a few days and buy a ticket. Is this even physically possible?

A:

It depends which way you want to go. Into the future? Maybe. Into the past? Almost definitely not. It's all thanks to a system called "causality" ....

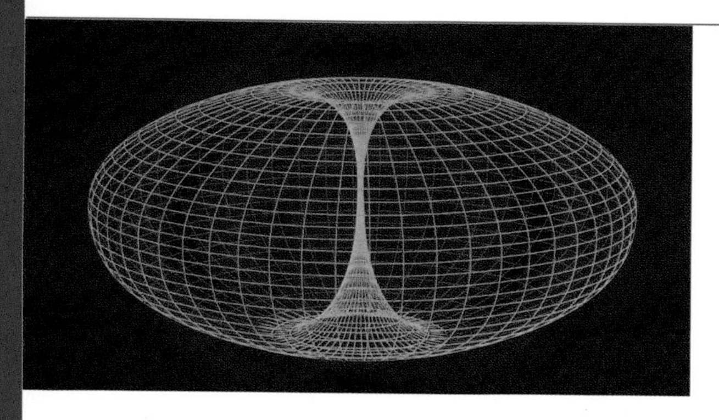

One of the most amazing things about the human brain is its ability to think up quite simple questions that, on further examination, turn out to be incredibly complex and difficult to answer.

It's very easy for a science-fiction writer to think up a time-travel story. The hero can step into a machine that travels in time in the same way a car travels in space. Push a lever to travel into the future, pull a different lever to go into the past. Easy!

When physicists start to investigate and do experiments on whether something like this is really possible, the math gets very complicated. Concepts like wormholes and event horizons and geodesics and worldlines get thrown about and chalkboards fill up with difficult formulae.

At our current level of understanding, it looks like time travel is maybe possible according to certain interpretations of Albert Einstein's Theory of General Relativity. But the act of building a real time machine might, bizarrely, cause the universe to destroy it instantly.

The physicist Stephen Hawking calls this the "chronology protection conjecture"—this idea that the fundamental laws of nature prevent any observer from being able to travel backwards in time. The math is hardcore, but in a nutshell it says that while you could theoretically open a wormhole to the past, the energy levels at the opening of the wormhole would quickly reach a point where the wormhole collapses again. Creating a wormhole also, paradoxically, destroys it. Confusing? Welcome to quantum physics!

Time travel to the future is a different matter, because this doesn't necessarily violate the principles of causality.

Causality is a hugely important concept that underpins almost all of physics. It's pretty simple: it just says that one thing causes another thing to happen. If you throw a ball, the *cause* of you accelerating the ball with your arm has the *effect* of the ball flying off across the park. What can't happen is the ball flying backwards across the park can't *cause* your arm to accelerate it.

However, there would be no problem for causality if you threw the ball and then instantly traveled forward in time to the point where the ball had been lying in the grass for a week. The effect of you picking up the ball would still come after the cause of you throwing the ball into the grass.

Time travel into the future is really just about you not experiencing the time between now and, say, next year. For you, time travel could happen if you went into a coma or some kind of suspended animation. Does this really count, though? Your body still moves at a normal rate through time.

Again, some interpretations of the Theory of General Relativity suggest it might be possible to skip forward through time, but whether these abstract mathematical concepts can be turned into a real time machine remains a challenge, ironically, for the future.

Wormhole Through Space

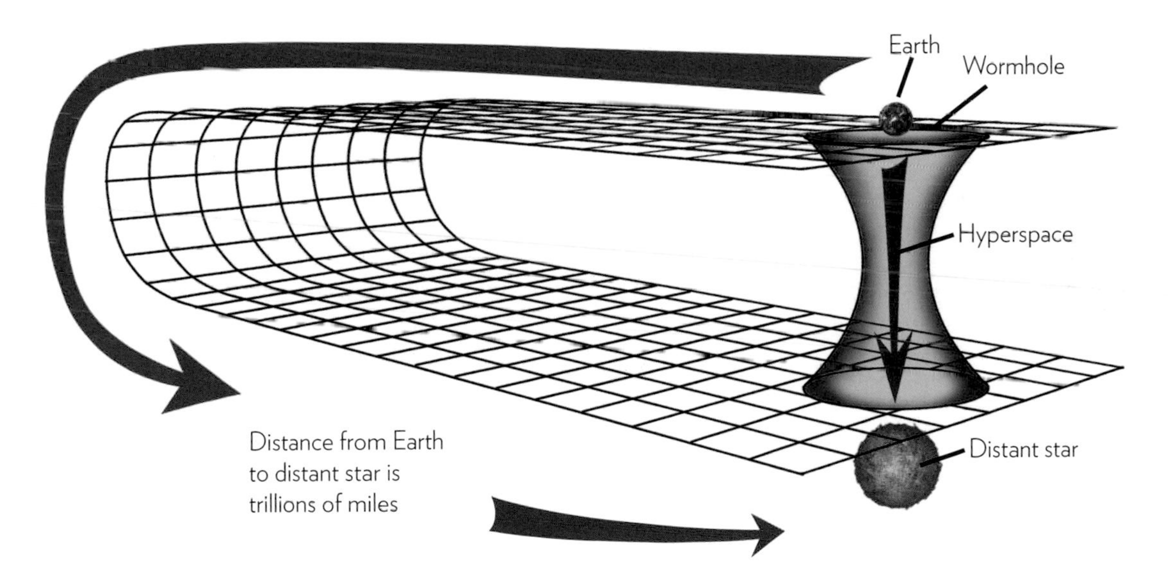

Q:

What's the big deal with the "Uncertainty Principle"?

There's a concept in physics that says we can't simultaneously know both the position and momentum of a subatomic particle. What does this mean, and why is it so important?

A:

At very tiny scales, matter behaves very strangely. Instead of knowing exactly where, say, an electron is, we have to make an educated guess. Yet weirdly, the guess can affect the real answer ....

Those pesky atoms. The universe was much simpler when we thought, before the twentieth century, that atoms were the smallest possible unit of matter. Then we had to go and discover things like protons, neutrons, electrons, protons, and quarks.

These subatomic particles are so small they don't behave in the same way as big matter (cats, cars, milkshakes, etc.)—even though big stuff is made up of subatomic particles.

When you or I get hit by a beam of light, the individual photons in that beam are so tiny compared to us that the physical effect of any single photon is virtually nothing.

But compared to an electron, a photon is actually quite sizeable. If you want to see an electron by bouncing light off it, the light can actually affect the electron. In fact, an electron is too small and weird for us to "see" it using light, and we have to invent particle accelerators and other complicated machines.

And sure enough, physicists did this and discovered something very strange. Whenever they pinned down the exact location of an electron, they couldn't then tell how much momentum it had—in other words, they couldn't see how much energy it would take to change the electron's speed. But on the other hand, if they measured how much momentum the electron had, they were unable to then tell exactly *where* the electron was!

If you think of an electron as a really tiny ball bearing, this doesn't make much sense. But even though we draw electrons as little spheres, that's not actually what they're really like. At this scale, matter doesn't have the same sorts of shapes it does in the big world. Matter behaves more like a wave—it exists in a sort of fuzzy cloud of possible locations.

There's a concept in physics similar to (and underpinned by) the Uncertainty Principle called the Observer Effect. This says that if you look at a subatomic particle, like an electron, you will affect its position or momentum. Just looking at it changes it. How is this possible?

Imagine that you looked at objects by throwing a football at them, and then catching the football when it bounced back. You could figure out how far away an object like, say, a skateboard was based on how long it took the football to come back to you.

But a football is actually heavy enough that when it hits the skateboard, it makes the skateboard roll away. Now, when you catch the ball again, you know where the skateboard *was* when you threw the ball, but you don't know where it is *now*. That's the Observer Effect.

In the late 1920s, physicists were worried that these effects were just being caused by the equipment they were using. But a physicist named Werner Heisenberg showed that the Uncertainty Principle was part and parcel of any wave-like system.

People often get the Observer Effect and the Uncertainty Principle mixed up. They are closely related, but the bottom line is that subatomic particles are weird.

How atoms are usually depicted

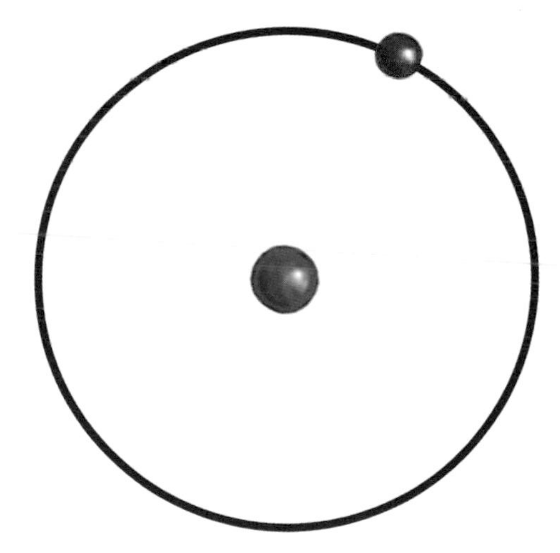

Hydrogen

A more realistic view of an atom

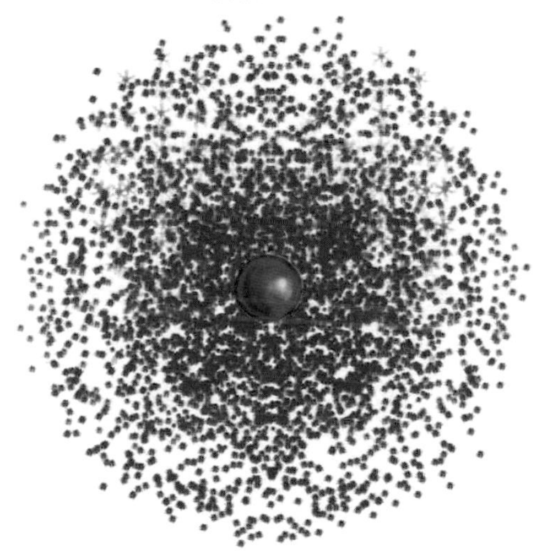

The whereabouts of the electron is somewhere in the cloud

Q:

How can a 10,000-ton boat float, while a 10-ton truck sinks?

Boats float because they're hollow, right? But why does this make it possible for a cruise liner to weigh so much more than, say, an airliner, which would sink?

A:

How heavy an object is overall matters less than how much water it has to move out of the way to sink. This "density" is what makes things float.

Very light things, like feathers, float easily. Very heavy things like bricks tend to sink. But this isn't a useful rule, because even the smallest boat weighs much more than a single brick. Even some hollow things, like steel boxes, will sink. So how does anything float at all?

Whether or not something floats depends on how dense it is compared to water. That is, how much it weighs per square inch, rather than how much it weighs overall.

The ancient Greeks were among the first to describe how this works. They observed that if you put something into a bowl of water, it takes up space that the water was previously occupying.

While you can force something underwater with your hand, if you want something to sink by itself it has to displace an amount of water that equals not its size, but its weight. A lead ball-bearing weighing five pounds will displace more water than a steel box weighing two pounds, even if the box is physically much bigger than the ball-bearing.

To get something to float, we just need to design it so it displaces its weight in water before that amount of water equals its size. Different objects have different densities, and if a boat's density is sufficiently low—usually thanks to hollow compartments filled with air—it will push down on the water at a rate that lets it displace its weight before the edge of the hull is underwater.

physics

We can control how low a boat sits in the water by increasing its overall density. Submarines flood their empty compartments with dense seawater. It doesn't make the submarine *that* much heavier, but it does make it quite a bit denser, so it moves lower into the water. To float again, the submarine pumps out the water using compressed air. Density drops, and the boat rises.

A regular delivery truck is not designed with low density in mind. It sits on four wheels and supports a large amount of weight at four small points touching the ground. The effective density of the wheels—with 10 tons of truck pressing down on them—is really high. Tons per square inch, in fact!

So if you drive a truck off a pier, the wheels will fall quickly below the surface but won't displace much water. The chassis, too, will be very dense, and that also will sink. However, depending on the truck, if there is a large empty compartment on the back, it might be light enough to overcome the weight of the wheels and chassis and keep the truck at least partly afloat. Until the compartment floods ....

An object like a truck or a boat can have very dense parts, but the whole thing adds together—which is why you can drive a truck onto a boat. If the combined weight of the boat and the truck displaces enough water before gravity can pull the whole thing under, the boat won't sink.

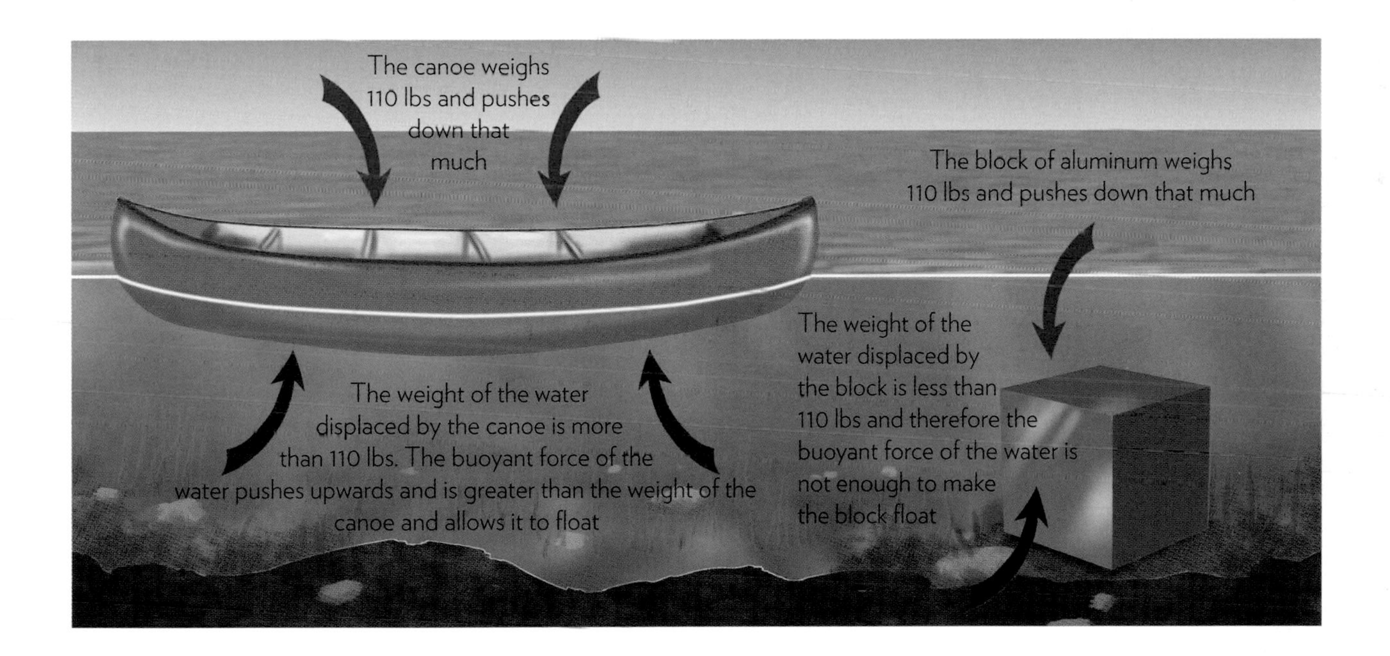

Q:

Every time we launch a rocket into space, does it affect the spin of the Earth?

Since every action involves an opposite reaction, does shooting a rocket off from the Earth actually push back on the Earth's rotation and slow the planet down slightly?

A:

Yes! But by such a small amount, it gets outweighed by the many advantages of using the spin of the Earth to give a rocket a boost into orbit ....

Launching a rocket is pretty simple: all you have to do is accelerate the spacecraft to 17,500 mph (28,164km/h) to enable it to reach an orbit 100 miles (161km) above the Earth. There, the atmosphere becomes so thin the spacecraft can "fall" around the curve of the Earth forever and never hit the ground. Couldn't be easier! Well, except for all the fuel and expense and red tape ....

This is how our satellite and space station orbits work: a few hundred miles up, gravity is still 95 percent as strong as down on the surface. So our stuff has to be accelerated to a certain speed and pushed up to a certain height.

When we launch a rocket in the same direction as the Earth is spinning—that is, from west to east—the rocket gets a boost to its speed. It's like throwing a ball forward while skateboarding. The ball will actually be going faster than if you'd thrown it while standing on the sidewalk.

NASA and other space agencies can save 5 to 8 percent in fuel by using the spin of the Earth to launch rockets. But there is a very small downside.

When you throw a ball, the ball takes energy from your arm and pushes back on you, even as it hurtles away. The best way to experience this is to set that skateboard on a very smooth floor, stand on it, and throw a dodgeball away from you with both arms. You'll roll backwards, at a speed determined by how heavy the ball is and how hard you throw it.

This happens to the Earth when we launch a rocket. The immense thrust of the rocket pushes against the ground, and so against the Earth. Back in the gym, if our skateboard was rolling forward slowly when we threw the ball, it would be slowed down as the ball left our hands. This happens to the Earth every time we launch west to east—the rocket pushes back and slows the rotation slightly.

By how much? Well, it's the tiniest fraction of the tiniest fraction of an inch. That's because the Earth is trillions of times heavier than the rocket. So this is one of those interesting physics points, rather than something to cause us to ban rocket launches.

Indeed, the advantages of using the movement and gravity of planets to launch spacecraft are massive. Planets are huge reserves of energy we can tap into, just by flying close to them. Several probes have used the massive gravity of Jupiter to slingshot out of the Solar System—even though doing this slows Jupiter down very slightly, too.

One of the most impressive boosts comes from just launching a spacecraft when the Earth is at a point in its orbit where it's moving *toward* the destination—say, Mars—rather than moving away from it. When we launch at the right time, our spacecraft can get a speed boost of over 100,000 mph (160,900km/h)!

This is why NASA talks about "launch windows"—the most energy-efficient times that only occur for a few days, and sometimes even hours, and might not come again for years or even decades.

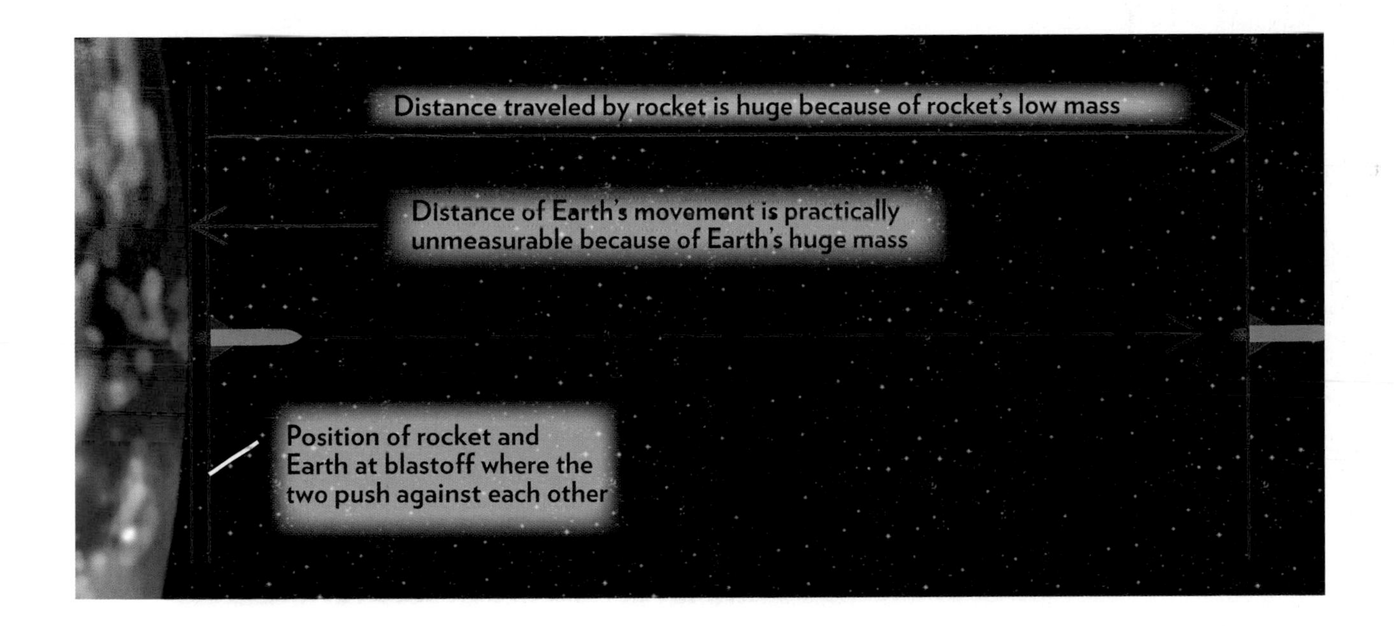

Q:

Why do atomic clocks that go up to the International Space Station appear to run slower in space?

Atomic clocks are supposed to be super-accurate, but experiments show that if you send one into orbit and compare it to one in the lab, it will appear to lose time. Why does this happen?

A:

The farther and faster you travel away from Earth, the slower time runs for you—at least, from the perspective of someone still on Earth. Sounds crazy, but it's explained by Einstein's theories of relativity.

One of Einstein's most important statements in his theories of relativity is that nothing in the universe can travel faster than the speed of light.

This has some weird implications. Like, what if you were in a long spaceship traveling at light speed, and you ran from the back of the spaceship to the front? Speed of spaceship (light speed) plus your speed running (a few miles per hour) should equal speed of light + few miles per hour.

Not according to Einstein and relativity. To stop the light speed limit being broken, time itself changes speed. To someone on Earth looking at you in the spaceship (they have a really good telescope and the spaceship is transparent for some reason), you will only be running at light speed, but it will take you longer to get to the front of the ship.

This is only seen by the person observing you. To you, time runs at a normal speed. Unless you look back at Earth—then you see the astronomer with his telescope, but it looks to you like *his* time has slowed down, too.

Wait—*both* people claim the other person's time has slowed down? That doesn't make sense! But really, it's a question of perspective. For a similar idea, stand 10 feet (3m) from a friend and hold your hand out. From your perspective, your friend appears to have shrunk so he takes up the same amount of space in your field of vision as your hand. But at the same time, if he holds his hand out, your friend will say *you've* shrunk to the same size as *his* hand. Crazy paradox?

This is what Einstein meant by relativity. There's no correct or special place in the universe from which to observe everything. Every observation is relative. To your friend, *you're* farther away. To you, your *friend* is farther away.

This might all sound like so much theory and fun thought exercises, but it has real implications for our everyday life. The idea that time flows slower for an object that's moving very quickly compared to you is something you might encounter every day through your smartphone's satellite navigation system.

The GPS satellites that give our sat nav systems their coordinates orbit 12,000 miles (19,300km) above the Earth, and they move at about 8,700 mph (14,000km/h). This is high enough and fast enough for their onboard clocks to run slightly slower (by our reckoning) than an equivalent clock on Earth.

The software in your sat nav (and on board the satellites) knows this and corrects for it, adjusting the time by a few fractions of a microsecond. Without this correction, the location calculated by your GPS receiver would get more and more inaccurate each day, eventually sending you miles off course.

To confirm this theory, scientists have flown atomic clocks in high-altitude, high-speed aircraft and on the International Space Station—which orbits at over 17,000 mph (27,359km/h). Sure enough, those clocks tick slower than an identical clock back home in the lab. Einstein's theory in action.

For a person on Earth, the clock in the passing rocket appears to run slower than the clock on Earth

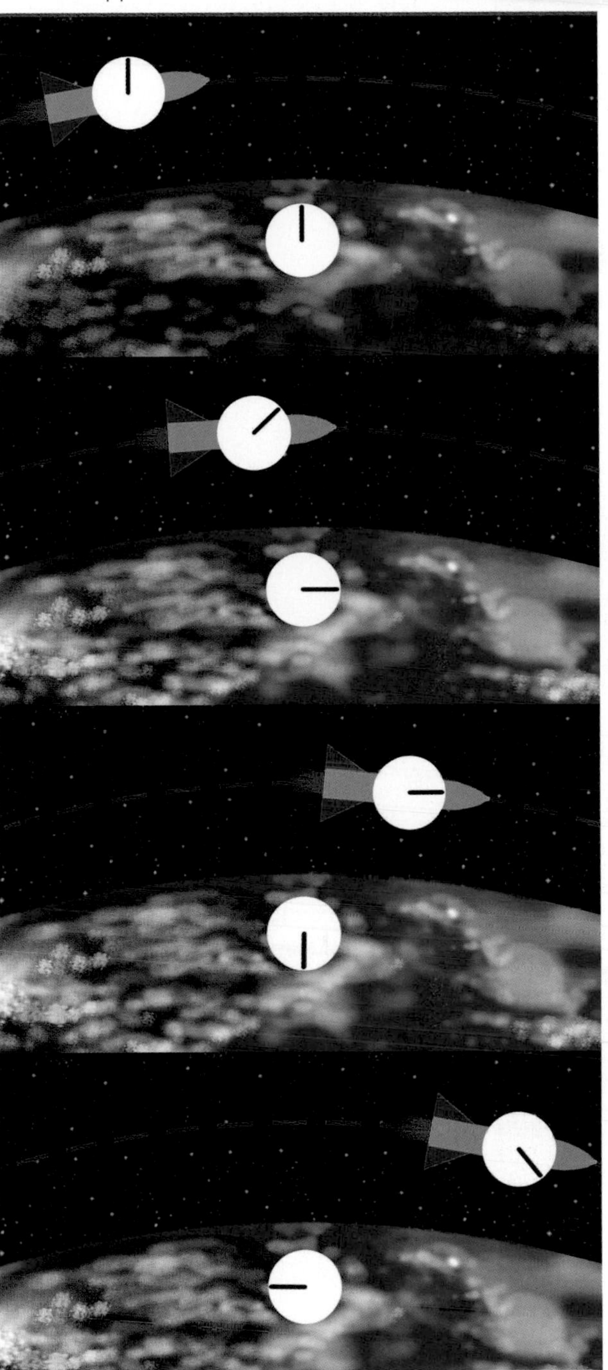

Q:

Could the Egyptians *really* have built the pyramids all by themselves?

One of the most popular conspiracy theories says the Great Pyramid and others of similar size are too big and too perfect for pre-industrial humans to have built. But could they have?

A:

Yes. Today's supercomputers can simulate how it could have been done. But the exact method remains up for debate ....

Modern humans, with all their fancy technology, have a tendency to be a bit patronizing about ancient civilizations. But since we're separated from them by just 4,500 years (the Great Pyramid was finished in 2560 B.C.), they were every bit as smart as us. They just didn't have jet airliners or electricity. Or wheels.

Egypt had a real thing for pyramids and built hundreds of them over several millennia. The Great Pyramid of Giza is the biggest and the most famous—it was already old when the Greeks and Romans visited. It's the only one of the classical Seven Wonders of the World still standing.

The original height was 481 feet (146.6m), but today it's 455 feet (138.5m) tall and 756 feet (230.5m) across at the base. It probably weighs over five million tons.

Mysteriously, the pyramid only has three chambers in it, plus a bunch of even more mysterious narrow shafts. These have puzzled people throughout history, leading some to claim the builders must have had supernatural or even extra-terrestrial help.

But this vastly underestimates both human ingenuity and determination. And remember: if you took an MRI scan of the brain of the man who designed the Great Pyramid and the man who designed the Boeing 747 jetliner, there would be no significant biological difference. These were sophisticated modern humans, not cavemen.

First, the stone: There are dozens of quarries both near the pyramid and along the banks of the Nile River that show evidence of stone blocks being cut with harder stone tools. There are half-finished stones still there, at various stages of construction.

These stones would have been loaded onto barges and sailed up to the construction site. Local stone would have been placed on sleds and dragged—the Egyptians didn't have the wheel, but wheels wouldn't have been much good for moving huge stone blocks over soft desert sand anyway.

After this things get a bit hazy. Oddly, there are no hieroglyphs or inscriptions describing the exact building process—perhaps it was a trade secret of the architect. So several theories—backed by experiment and math—have been put forward.

The most popular explanation is that along with the pyramid itself, the builders constructed huge ramps up the sides, upon which they dragged the stones. But since the top of the pyramid contains so little stone, and you'd still need a big ramp to get there, this seems inefficient.

A more radical theory is that the ramps are actually *inside* the pyramid and remain embedded in the structure. Even though this theory is less popular, thermal imaging of the Great Pyramid does indeed show these ramp-like structures.

This also explains the mysterious shafts. It's possible the architect was worried about the pyramid cracking under its own weight, so he had the shafts drilled and filled them with plaster. He then checked the plaster regularly for new cracks.

These explanations are less exciting than the idea of aliens visiting just to build a pyramid or two. But they do show what humans can achieve, and that limited technology is no impediment to accomplishing amazing things, if you've got the patience. Also slave labor. Slave labor helps, too.

Q:

When did pcople stop believing the Earth was flat?

Early humans had no idea they lived on a sphere in space, but thought they stood on a great flat disk. When did we finally realize the Earth is actually a sphere?

A:

About 600 B.C., give or take. The ancient Greeks used their theories of mathematics and careful observation. Though *proof* of a spherical Earth came much later ....

Unfortunately, there's a widespread belief out there that most humans thought the Earth was flat until Christopher Columbus sailed to the Caribbean in 1492. In fact, this idea was put about by people in the eighteenth century who thought people in the fifteenth century were stupid.

Humans came upon the idea that the Earth was round pretty much as soon as one civilization got big enough to have settlements spread across a large north-south distance. That civilization was ancient Greece. Since they had towns and cities up in Central Europe and quite far south into Africa, they could exchange observations.

What these early Greek observers saw was that the positions of the stars changed as they traveled north or south. The farther south they went, the higher the southern stars would rise each night. This only made sense if the surface of the Earth curved away in every direction—if it was a sphere.

This also neatly explained why the shadow of the Earth on the Moon during a lunar eclipse appeared to be round.

Famous Greek philosophers like Aristotle and Plato taught their students that the Earth was round. Mathematicians like Pythagoras assumed it was round.

While the *theory* of a round Earth has existed for more than 2,500 years, an actual solid, no-nonsense proof we live on a sphere is much more modern—only 500-odd years old.

While humans circumnavigated Africa and traveled back and forth from Europe to China and eventually even discovered the Americas, all this proved for sure was the surface of the Earth was curved in every direction—it still might not have been a sphere. It could have been a bowl or something weirder.

Not until Magellan's expedition finished its circumnavigation of the globe in 1521 did we have absolute, undeniable proof that we lived on a "closed surface," or a sphere where you can go in a single direction and return to where you started (at least, you can if you have an airplane).

The first photograph of the Earth from space was taken in 1959 by NASA's Explorer VI satellite. It looks like a grey blur, but you can sort of tell it's round.

The most famous single image of Earth is the so-called Blue Marble. It was taken in 1972 by the crew of Apollo 17 from 28,000 miles (45,000km) away and is a "full view" picture—the sun was directly behind the spacecraft so the Earth wasn't shadowed at all.

It's probably the best historical artifact for proving to someone that the Earth is a sphere. Unlike other pictures that are made of stitched-together satellite shots, the Blue Marble was taken by a single camera—a Hasselblad 70mm with an 80mm lens. Point and shoot. But the astronauts were upside down, with the top of the camera pointing toward the South Pole. It's lucky photos are easy to flip!

Knowing the distance between Syene and Alexandria and using a well in Syene that cast no shadow at noon and the shadow cast in Alexandria, to the north, at the same time of day, Eratosthenes (276 - 195 B.C.) calculated the approximate circumference of the Earth.

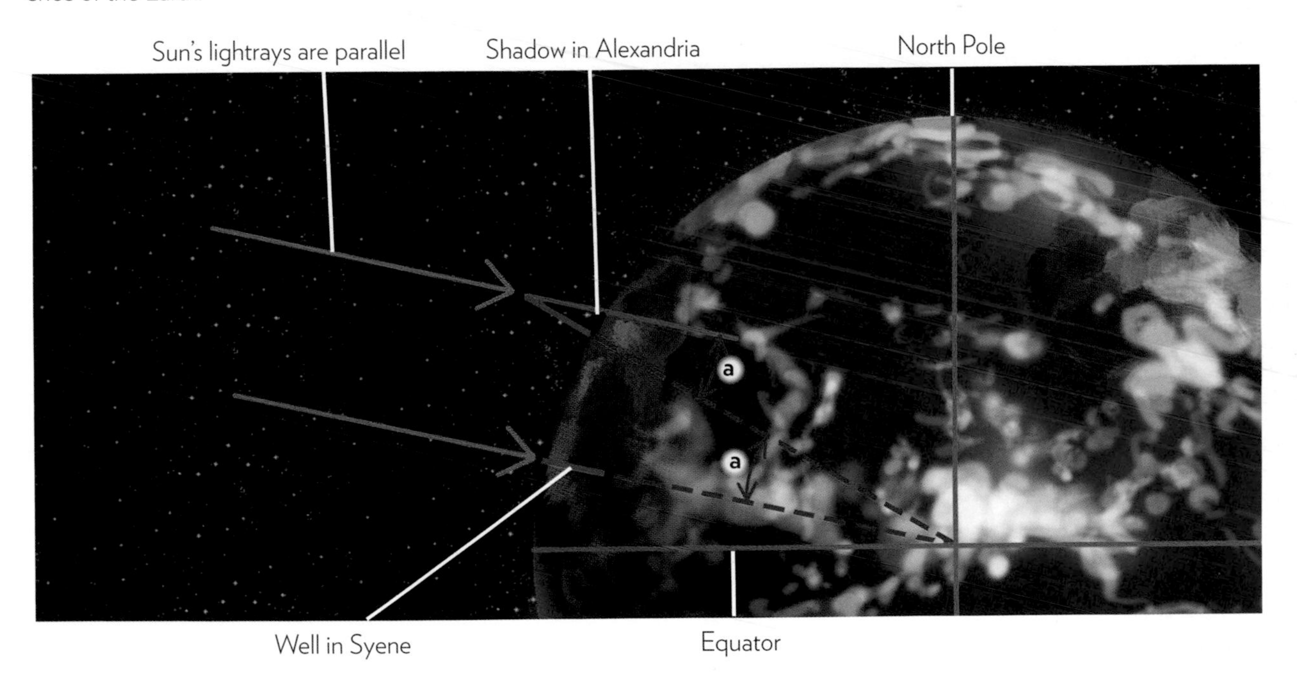

Q:

How can I be sure the Earth orbits the Sun?

Is there an easy way to prove the Earth orbits the Sun without needing a spacecraft or relying on a textbook?

A:

An easy way? Yes. A quick way? Not really. Proving the orbit is as easy as looking up at the stars. Every night. For a whole year ....

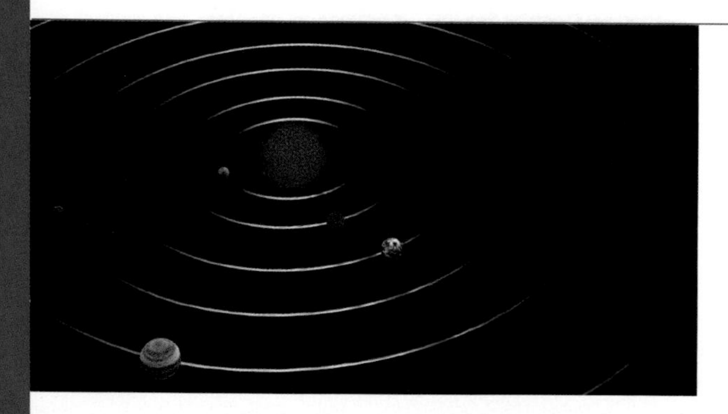

There are a lot of so-called universal truths that we take for granted. Many of us are happy to believe the Earth is round because we can jump in an airplane and keep flying west (with a few stopovers) until we get home again. Proof!

But what about the Earth orbiting the Sun? We're taught this, we accept it, but what if we had to prove it to, say, a bunch of indigenous tribesmen in the Amazon? Or to a crazy person? Is it possible?

You can indeed see that we orbit the Sun with your own, unaided eye. All you need is a lot of patience and somewhere with a good view of the stars.

What you will notice is that, over the course of the year, the patterns of stars in the sky will change slightly. Some constellations will disappear from one side of the sky, and others will appear from the opposite side. Then those stars disappear and the original stars return. In other words, the whole of the heavens rotates over the span of a year.

Unfortunately, this doesn't by itself prove we orbit the Sun. What *could* be happening (and what humans believed for many thousands of years) is that the whole sky could be rotating around the Earth.

To prove the Sun is the center of the Solar System requires another year of even more careful observation.

Even though the stars look like they're set on the inside of a giant bowl, they are all different distances from Earth. Some are much closer than others.

If you pick two stars that are close together, odds are one of them is actually much closer to Earth than the other. Keep your eye on these stars over the next year—you'll see the farther star slowly cross behind the closer star until it appears on the opposite side.

This is because the Earth is moving around to different positions in its orbit. Let's imagine looking down on the Solar System as if it were a giant clock. And we'll say the two stars are at the 12 o'clock position, one above the other.

When the Earth is at 3 o'clock, the farther star will be visible and appear to be to the right of the closer star. Later in the year, when the Earth is at the 9 o'clock position, the farther star will appear to be to the *left* of the nearer star. And when the Earth is at 12 o'clock, the farther star may not be visible at all.

The reality of doing this "parallax" observation is a little more complicated, but the principle is there. By noting carefully how the positions of the stars in the sky change, it's very difficult to come up with any explanation other than the Earth orbiting a central point. In this case, the Sun.

Closer stars appear to move very slightly relative to other stars over one year's time

Q:

Why are tornadoes only common in some areas?

Tornadoes are terrifying, but at least if I keep away from the American Midwest I probably won't even see one. Why are they only encountered in a few places?

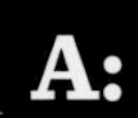

A:

Tornadoes need very specific conditions to form. The midwest United States has almost perfect tornado-making geography, but there are a few other global hotspots, too.

A tornado is a special kind of storm that produces a very narrow and very intense column of rotating air. As anyone who has ever lived in America's Midwest knows, tornadoes can be extremely destructive.

The atmospheric processes that form tornadoes aren't fully understood yet, but our knowledge is improving. We do know that for a tornado to form, warm, wet air has to collide with cold, dry air, and if it does this over flat land with no mountains in the way, tornadoes can develop.

There are only a few places on Earth where this happens regularly, and they all have something in common. They're halfway between the cold arctic (or antarctic) regions and the equator, and they also have wide, flat areas and—this is key—no mountain ranges running east-to-west to block the movement of air along the ground.

When big thunderstorms become very powerful, they can form "mesocyclones" several miles up in the atmosphere. In the right conditions, a process called a "rear flank downdraft" will pull a mesocyclone toward the ground, where it will form a tornado. Spinning air high in the clouds isn't very dangerous, but when it touches the ground, all hell breaks loose. Also your house—that can break loose, too.

America's Midwest is one of the most perfect places for tornadoes to form, because there are mountain ranges that run north to south on both sides of wide open plains. These ranges funnel cold air flowing down from Canada where it smashes into hot, wet air from the Gulf of Mexico. The Midwest can see up to 1,200 tornadoes a year, though these are spread over quite a wide area.

If you take area into account, the place with the most tornadoes per square mile is actually Holland! But these are less powerful than tornadoes in the United States, though they can still cause damage. The United Kingdom also has a lot of tornado activity, though again they are much smaller and less destructive than U.S. tornadoes, so they don't get the same amount of media coverage. Many European tornadoes form off the coast and are called "waterspouts." These are normally less dangerous than a tornado on land.

The country that loses the most people to tornadoes each year is Bangladesh. Again, the tornadoes there are not as powerful as in the United States, but the population density is extremely high and there's not much education about how to defend from severe weather.

Australia is more or less the opposite: it has many quite powerful tornadoes per year, but these mostly occur out in desert regions where nobody lives. The majority of Australia's big tornadoes go unreported and unnoticed by anyone except climatologists looking at satellite images. Though in 1970, a tornado to rival some of the United States' most powerful ripped through a forest near the town of Bulahdelah.

Tornadoes are also common in South Africa, New Zealand, and the bottom half of South America.

World Tornado Map

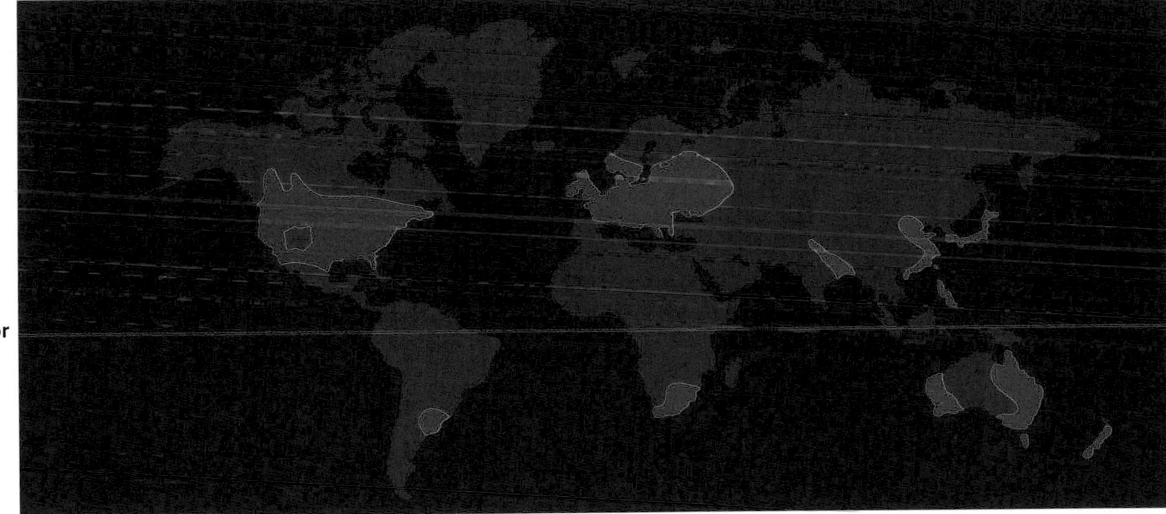

Q:

Why don't the filaments in cheap light bulbs last forever?

Do light bulb makers deliberately make bulbs that don't last very long? Is it a conspiracy to get us to buy more light bulbs?

A:

There's no conspiracy, just a limitation to bulb technology—and for manufacturers, it's a happy one since it does lead to more bulb sales. The secret? Rust!

It's kind of amazing that you can light your home with a 100-watt bulb for a couple of bucks. But the downside is these bulbs burn out within a few months. The cheaper the bulb, the shorter the life span.

Is it a conspiracy? Aren't humans smart enough to come up with a light source that's as bright as a bulb but doesn't pop? Well, yes, of course we are: you can buy halogen lamps that last for thousands of hours or fluorescent tubes and bulbs. Today, the latest thing is the light-emitting diode or LED, which measures its life in *years*.

All these technologies have one thing in common: compared to a regular light bulb, they're very expensive. LEDs, right now, cost 10 to 20 times more than an equivalent bulb.

The familiar incandescent light uses a tungsten filament in a bulb full of an inert gas like nitrogen. The filament is basically a very thin wire, often coiled into a tiny spring shape.

It has a useful property in that when you pass an electric current through it, the filament glows and puts out visible light.

The filament isn't burning like a wick on a candle, but rather the atoms of tungsten are resisting the flow of electrons from the socket. The tungsten disposes of extra energy by putting out light. If you change the amount of resistance the bulb has, the amount of light changes, too.

It turns out this process comes with a cost: whenever the filament is glowing, a few atoms of tungsten get evaporated, making the filament very slightly thinner. What's more, air will leak in around the edges of the bulb and react with the filament, causing it to rust.

Since the filament rusts and evaporates unevenly, this creates weak points in the wire. Where the wire is weaker, its electrical resistance is different. Electrons rush to this area and can create a hot spot.

Eventually, the stress is too much and the filament snaps, often with a loud pop! Older bulbs could even shatter, but today's glass is stronger.

The reason this usually happens when you turn the light on or off is that when the flow of electricity changes through the wire, that's when hotspots and stresses occur. If the bulb is just left burning, it will last longer (on average, anyway).

Still worried it's a scam? A standard tungsten light bulb is a compromise between a bulb that puts out lots of light, versus a bulb that will last a long time, versus a bulb that doesn't cost very much. Manufacturers could double the thickness of the filament, but this would add to the cost and affect both the intensity and also the color of the light.

Of all the undesirable possibilities, having bulbs that blow is considered the least worst. And yes, it does mean they can sell you a new one.

Some points become weaker than others and resistance, which creates the light, becomes greater and, therefore, hotter at those points until the bulb burns out

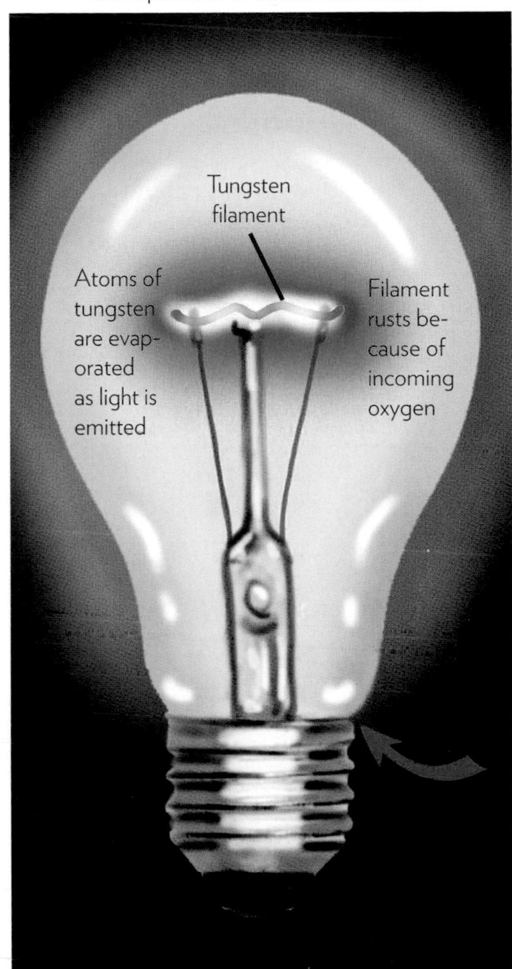

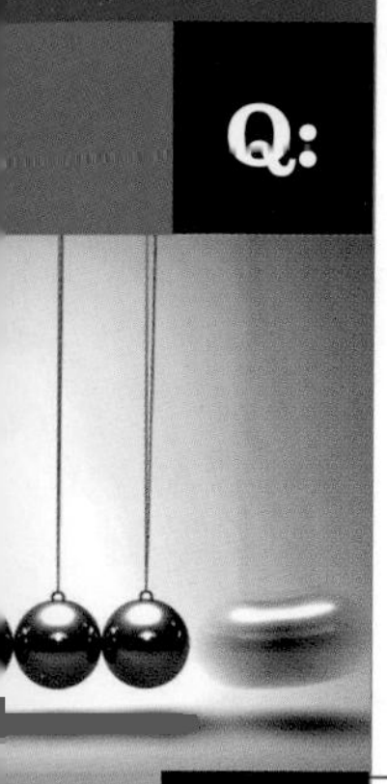

Q:

How do we see colors?

Our eyes detect light particles (photons) hitting our retinas. But how do they tell the difference between the various colors of light?

A:

Our retinas have special cells that respond to particular wavelengths of light. These signals combine to create color. Humans have good color vision, but some animals beat us hands down.

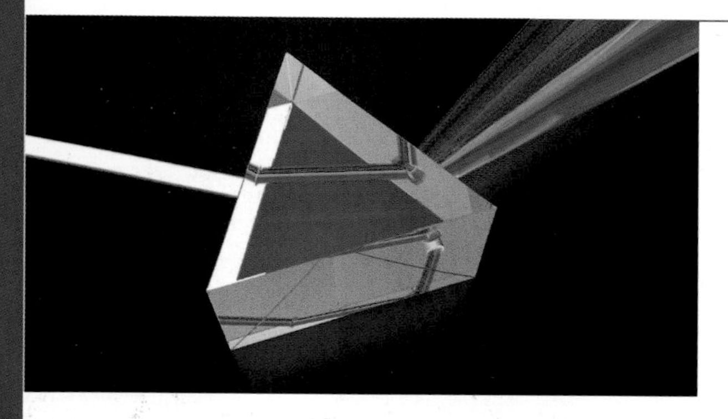

Most of the light we use to see objects here on Earth comes from the Sun. The Sun puts out light at lots of different frequencies—which means the light vibrates up and down at different rates, like waves rolling in to a beach.

We call the whole amount of light coming from the Sun "white light," because when we shine it on a white object, well, that object looks white! The light itself contains all the colors we can see.

Scientists show this with crystal prisms that can split light into a rainbow of different colors. It was Isaac Newton who explained that these colors appear because they travel through the prism at different speeds.

In other words, the color "blue" is just light with a frequency a little bit faster than "red" light. The familiar rainbow is a stack of light waves with the slowest on one side (red) and the fastest on the other (violet).

Light can have a color, of course, if it comes from a source like a flame, a light bulb, or a glow stick. This light only moves at certain frequencies, giving it a color like orange or yellow or pale green.

physics

Our eyes have evolved to detect these different frequencies of light. Around seven million special cone cells in each of our retinas pick up certain frequencies, and each individual cone sends a signal to our brain, which combines all the signals to create a color.

For example, a ripe yellow lemon absorbs all the blue light that hits it and reflects lots of red and green. Our cone cells pick up the red and the green, and our brain combines it into the color we call yellow. A clear sky reflects a lot of blue light, and our cones pick this up, too.

We have three types of cone cells: one for red, one for blue, and one for green light. From these three types of color receptors, with added information about shading and tone, come the 10 million colors humans can see.

Humans have very good color vision, which is unusual in mammals. Dogs and cats see fewer colors because they have fewer cone cells. But as a trade-off, they have much better low-light vision than we do, because they have more "rod" cells. These rods detect the strength of the light hitting them—they're what enable us to tell the difference between bright and dim light.

Other animals, such as birds, have better color vision than humans. They have four kinds of cone cells instead of just three, and this allows them to see light at ultraviolet frequencies. This is very high-frequency light that we usually think of as the stuff that gives us sunburn.

There are even animals with super color vision. A crustacean called a mantis shrimp has an incredible 16 different kinds of color receptors—12 for color detection and 4 for color filtering. These ocean dwellers only grow to about 12 inches (30.5cm) long, but they are able to see billions of different colors.

We have three types of cone cells covering the retina of the eye: one for red, green, and blue light. From these three colors, and by adding in information about shading and tone, come the ten million colors humans can see.

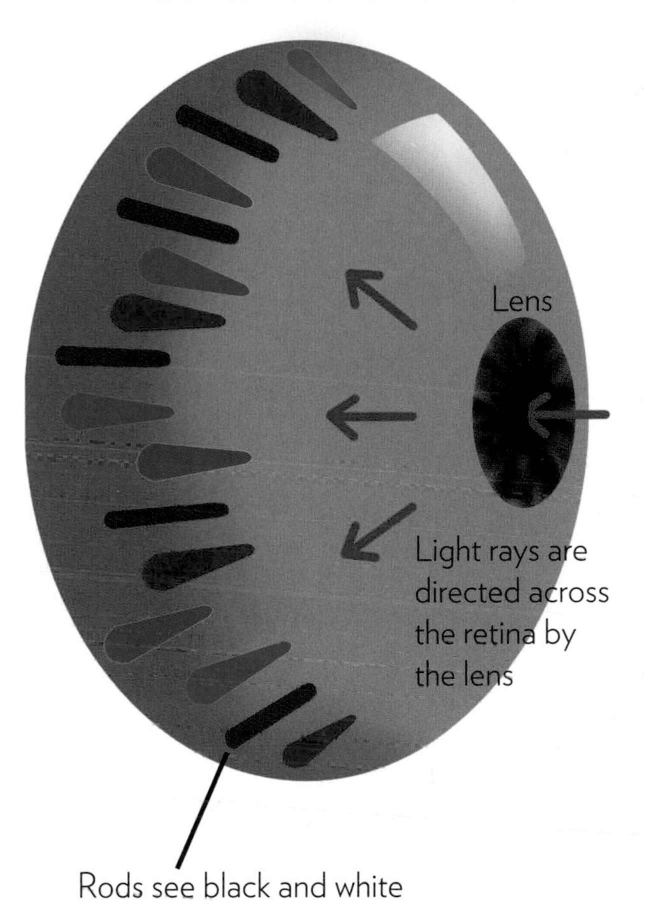

Q:

Why do magnets stick together?

Certain metals are magnetic and can attract other metals, but what makes magnets stick together so strongly?

A:

Magnetic fields have a direction, and when two magnetic fields point the same way and are close together, they reinforce each other.

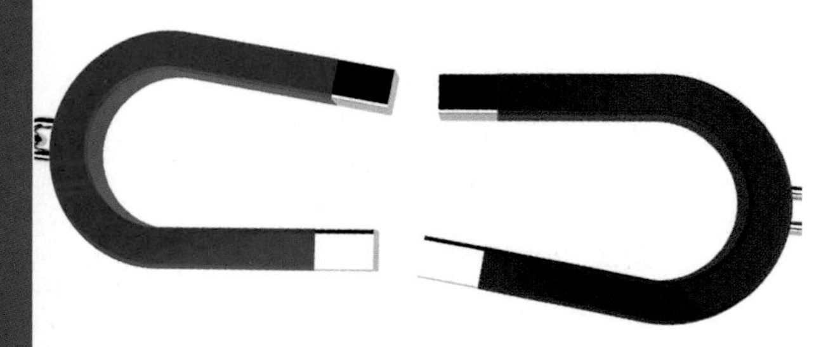

Every atom in the universe has some kind of magnetic field, created by the way its electrons are arranged around the nucleus. In a material that isn't magnetic, like plastic, these atomic fields all point in random directions, effectively cancelling out the plastic's overall magnetic field.

But in so-called "ferromagnetic" materials like pure iron, the magnetic fields line up much more closely, and this reinforces the field. The overall piece of metal is magnetic.

In some types of metal, the magnetic alignment inside is kind of loose or floppy. The metal won't work as a magnet all by itself, but if a strong magnetic field comes close, all the little magnetic fields inside the metal spring into life and point in the same direction—exactly like a bunch of tiny compasses all pointing north.

If you rub a piece of metal with a strong magnet, this can more permanently reinforce the metal's magnetic field, and it can become what we think of as a "magnet" in its own right.

When we use magnets, we are actually using special materials that have a very strong magnetic field. Some magnets, like those made of neodymium, are so powerful they can crush your fingers if you handle them incorrectly!

While magnets will stick to any magnetic material, they *really* stick to other magnets. Again, it's because when you align two magnetic fields, they want to join together. But this only works in one direction.

A magnet has lines of force that flow from the top to the bottom of the magnet—we call this "north pole" and "south pole." The south pole of a magnet will stick to the north pole of another magnet. If you try to stick a south pole to another south pole, the magnets will resist this. It feels kind of like a little cushion of nothingness. Small magnets can be forced together, but they will spring apart again as soon as you let go.

This force isn't made of anything—it's one of the fundamental forces of nature and is formed by the interaction of various quantum particles.

A magnet can lose its magnetism very, very slowly over time, but as rocks in the Earth's crust show, this can take millions of years. We can make new magnets and "recharge" magnets by realigning their magnetic fields.

The temporary way to do this is to simply put the magnet in a much stronger magnetic field, such as one made by an electrical generator. Or we can melt down the metal inside the magnet, align the atoms inside a magnetic field, and let it cool. This is a more permanent way of magnetizing something, and the way magnets are made in the first place.

Today, we use powerful neodymium magnets in everything from computer hard drives to massive turbines in power stations to the speakers in our earbuds. Some people even get magnets embedded in the bone of their jaw to help keep their dentures in!

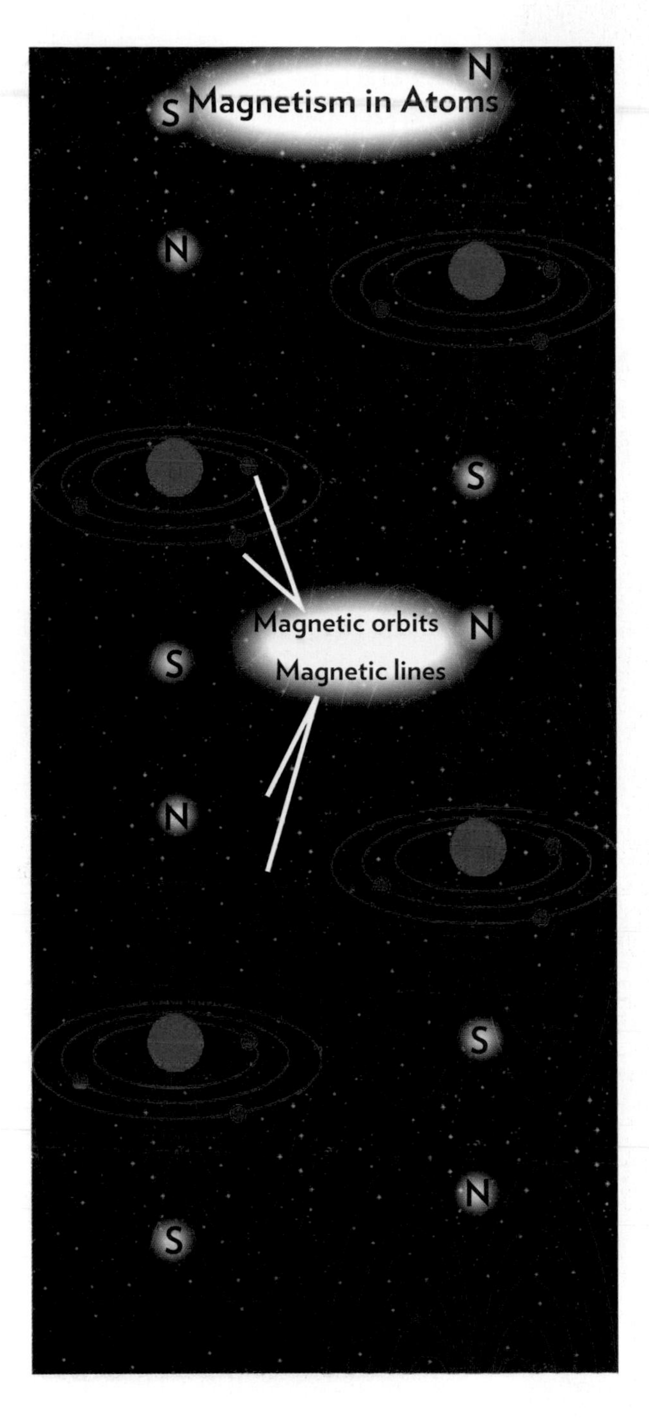

Q:

What would happen if the Sun collapsed into a black hole?

If our Sun turned into a black hole, would the Earth be sucked in and crushed? How would it affect the other planets? What exactly would happen?

A:

We would all freeze to death. The black hole version of the Sun wouldn't suck us in, because the gravity of the Sun wouldn't change, even if it did change into a black hole.

A black hole is one of the weirdest things in nature. These objects form when so much matter gets packed into a tiny space that it more or less punches a hole in the universe itself. The gravity near a black hole is so intense that not even light can escape from it—which is why we call them "black."

The sci-fi idea of a black hole sucking in everything that comes near it is technically correct, but the way novels and movies apply this idea is usually a bit wrong. For instance, a black hole can't pull you in from across the galaxy. You have to actually fly your spaceship within range of its gravity.

If the Sun did collapse into a black hole, it wouldn't get any heavier, it would just get smaller. Instead of being 864,000 miles (1,390,474km) across, the Sun would form what's called a "singularity." It would weigh the same but take up almost no space at all.

We wouldn't see this singularity, though. As you get closer to a black hole, gravity grows stronger and stronger. Eventually, gravity becomes so strong that light can't escape its pull. This means we can't see anything beyond that point, because the light can't get to us!

Physicists call this the "event horizon" of a black hole, and while we haven't yet taken a picture of one, it could look like a perfectly black, perfectly round circle. Or it could be hidden beneath bright streams and jets of radiation. But the size of the event horizon depends on the weight of the black hole itself.

The Sun's event horizon will be much smaller than our star's current diameter. And because all the planets are in stable orbits right now, the amount of gravitational pull the Sun exerts on them won't change. Their orbits will continue more or less as normal. Only the shape of the Sun will change, not its gravitational pull.

Of course, the sky will go dark and temperatures on Earth will plummet. There might be other effects, too. For instance, black holes may spew out lots of deadly radiation, and this could sterilize the surface. The force of the radiation might even be enough to "push" Earth off its orbit and send us careening out into space.

It also depends on whether the black hole starts spinning. If it does spin, the exact shape and various angles of its gravitational field could fluctuate. This, too, might disrupt the orbits of the planets, sending them off into deep space.

Because we haven't yet seen a black hole up close, we don't know for sure what would happen to a planet in orbit around one. It might even be that to form a black hole, the Sun first needs to explode violently, which could destroy the Earth early on.

The good news is that, based on our current models, our Sun isn't heavy enough to form a black hole. Instead, it will expand into a red giant ... which will most likely swallow the Earth. It will then contract into a white dwarf and remain as a slowly cooling cinder for the next trillion years.

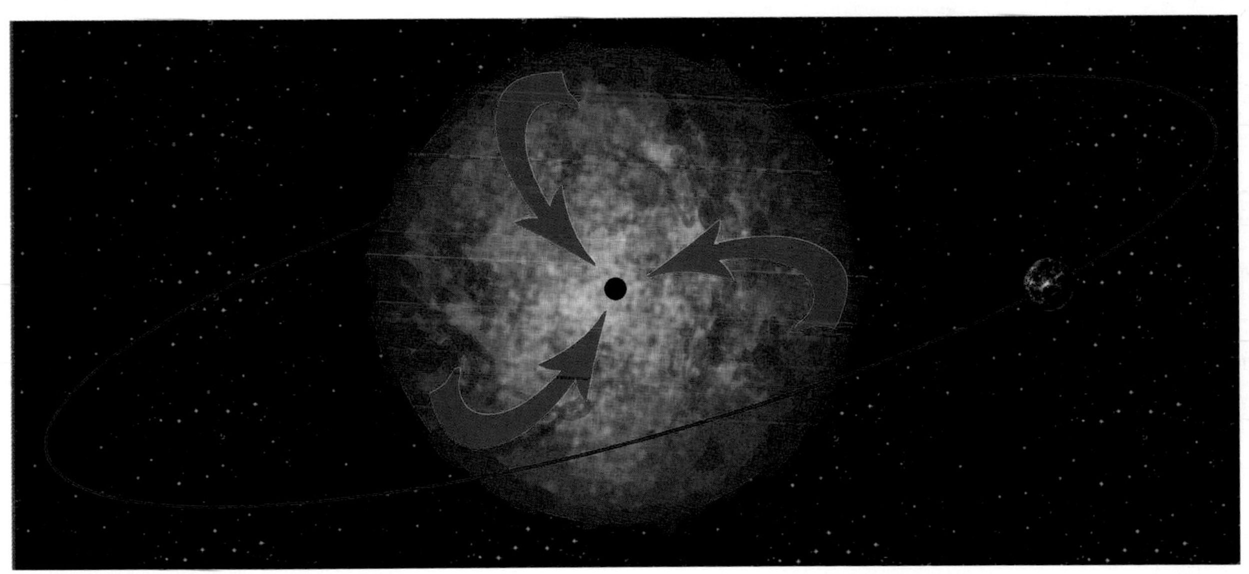

As a black hole, the gravitational pull of the Sun would remain unchanged

Q:

Why is a powerful electrical current so lethal?

Everyone knows electricity can kill, but how does this invisible force made of tiny electrons take down a person so quickly and so terribly?

A:

Electricity is so lethal it can kill you twice: once by burning your internal organs, and again by making your heart go haywire. But not all electricity is so deadly.

As useful as it is to power our machines and light our homes, electricity is also deadly to all life if it gets out of control. But how and why can it kill?

Electricity is essentially a flow of electrons from a positive region of charge to a negative region of charge. If we think of positive as the top of the mountain and negative as the ocean, then electricity is a river of electrons flowing down the mountain. Except this river flows at the speed of light and can also make your heart explode if you fall in.

Electricity isn't actually a fluid, but like water you can measure electricity based on how hard it flows (pressure in water, volts in electricity) and also how much of it flows (volume for water, amperes for electricity).

If someone squirts you with a water pistol, that's a small amount of water at high pressure, and it's harmless. And wading through a slow-flowing chest-deep river is also reasonably harmless—lots of water, but low pressure.

If you step into a raging flash flood, though, you'll be swept away and drowned. In the same way, electricity with both high volts and high amps is deadly. And here's why.

The first problem: burns. All matter has a property called "resistance," by which the atoms inside it will try to stop electricity flowing through. When electricity hits something, such as a wire, the electrons will sort of bunch up and force themselves through. Some of the electrons will let out energy, which transforms into heat.

physics

The human body has reasonable electrical resistance, but it also has lots of water. These two things combined mean that when a powerful jolt of electricity passes through the body, tissues and bones in the way will heat up. This heat is incredibly intense, enough to burn cells. All the water absorbs the heat and expands, and that does even more damage to our cells.

After a powerful electrical shock, the victim will have a hard, leathery entry wound and a puffy exit wound. Internal damage can be very severe, even bad enough to kill. But odds are this won't be the way an electrical shock does kill you. It's more likely the electricity will actually scramble your internal circuits.

Humans (and all life) run not just on oxygen and chemical energy, but also on electricity. We rely on electrical impulses from nerves to make our heart pump and to make every tiny valve and muscle in our body move.

When electricity enters the body, it seeks out areas of lower resistance, including parts of our heart. The massive surge overpowers the heart's natural control system and makes it start to quiver and twitch chaotically. This is called "fibrillation," and it can be fixed with another jolt of more controlled electricity—those shock pads you see in the movies.

Without immediate medical attention, this fibrillation starves your brain of oxygen and, sadly, you die.

As a general rule, an electrical current of more than about 70 milliamps is enough to send your heart into fibrillation, while a current of 1,000 milliamps is strong enough to burn.

Q:

Why do so many people survive being struck by lightning?

If a household power outlet can electrocute a person, how does anyone ever survive being hit by a gigantic bolt of lightning?

A:

Dumb luck, mostly. Lightning is very powerful, but it's brief. If it passes through you quickly enough, you might get away with nothing more than permanent brain and nerve damage ....

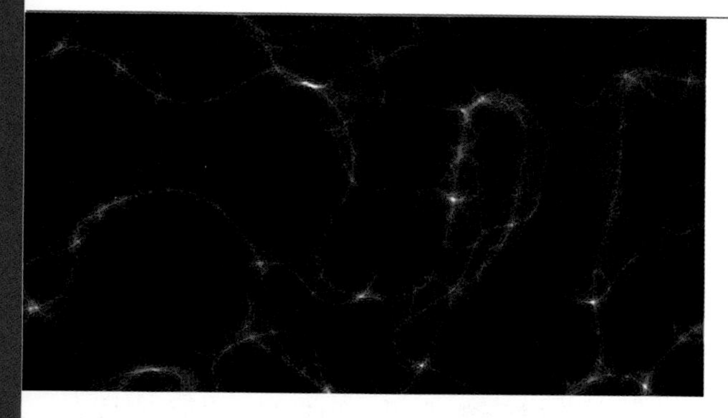

Most of the stories you hear about getting struck by lightning come from survivors—which makes sense, really. Of the 240,000 people hit or grazed by lightning strikes each year globally, about 24,000 are killed. The rest survive.

We don't want to understate the power of lightning. Your wall socket has 110 to 240 volts, depending on which country you live in. A bolt of lightning has a *trillion* volts. That's the power of the electricity, but what about the amount? That wall socket has 20 amps, and a lightning bolt has, well, 120,000 amps.

But when you foolishly stick a fork in your wall socket, the electricity gets conducted through the fork into your hand and keeps flowing until the safety device in your fuse box trips. If you have faulty wiring, you can end up connected to that electric charge for long, deadly seconds.

A lightning strike, on the other hand, is more like a near-instantaneous pulse of electricity. It's powerful but momentary. And that briefness is what gives you the chance to survive.

For a lightning strike to electrocute you, it has to move through your body on a path that hits your heart. There, the electricity will overwhelm your heart's natural electrical systems and send you into fibrillation. Instead of pumping, your heart will just quiver and twitch. Your blood won't flow and your brain will be starved of oxygen. If someone is on hand to offer CPR, you can survive.

But if the strike doesn't pass through your heart, you might be in luck. That is, if you think nerve damage and a lifetime of medical problems counts as luck. Survivors can have trouble forming new memories and problems with coordination, and suffer many other long-term effects.

Electricity, especially lightning, wants to make its way from a region of positive charge to a region of negative charge. The ultimate negative "sink" for electricity is the Earth itself. This is why lightning stabs toward the ground—it's seeking the Earth.

Electricity is most deadly when it flows. If you provide lightning with a path through to the Earth, you'll have lots of electricity flowing through you, and if that lasts for even a second, it will probably be fatal.

But if you somehow reduce the amount of time it takes the lightning bolt to "ground," perhaps by curling up into a ball on the ground to present a really tiny surface area, then you can get away with a lesser injury.

Hiding in a car or a shed with a metal roof and metal walls can help, too. But lightning is so powerful it can create shockwaves as it blasts apart the air, which can knock you down—a whole other way to get injured.

Many victims of lightning strikes aren't hit square-on by the bolt itself. They still get electrocuted by electricity arcing, or jumping through the air, but it's less powerful. So being missed by just a few feet can mean the difference between life and death.

Q:

Is wireless electrical power really possible?

Electricity is great, but the cords are a drag. Can't we get rid of them and have wireless power, the way we already have wireless communication and data?

A:

The answer is a qualified yes! Wireless electrical power is already available for some gadgets, but it only works at short range. Long-distance transmission of electricity is more problematic, though it might have been invented back in 1899.

If there's one thing that defines technology in the first decades of the twenty-first century, it's this: wireless. We have cellphones for wireless communication, and we have wi-fi for wireless data. Now where's our wireless power?

It does seem a little strange that we still have to plug in our smartphone every evening, connecting it to a source of electricity via a strand of metal. Surely science has come up with an alternative, a way to get the power to our gadgets without all those annoying wires?

Wireless charging is available right now on the latest smartphones and on humbler gadgets like electric toothbrushes. A toothbrush is probably the cheapest wireless electric device you can buy. It charges up when you set the toothbrush on a special cradle. The cradle is still connected to the wall outlet, but it uses electricity to power up an electromagnet inside. When you put your toothbrush on the cradle, the magnet stimulates a coil of wire inside the toothbrush into producing electricity. This is called "induction," and it's very handy because it means your toothbrush doesn't need metal contacts or a metal plug hole, which could get damaged by water in your bathroom.

Using this same system—a coil of wire and a magnet powered by electricity—it's now possible to charge some cellphones and gadgets in the same way. You just place the cellphone on a special mat or pad, and it charges.

Okay, this is kind of cheating. It's less wireless and more plug-free charging. You still need to remember to put your cellphone in a special place. What we really want is a system that constantly charges our phones while we're walking around. We want electricity sent into the battery in the same way data is sent into the phone.

This would have enormous implications for all sorts of things, especially cars. Electric cars that charge via wireless systems could have near-infinite range. It would be awesome!

Unfortunately, transmitting useful electrical power over long distances without wires—or rather, by using the air as the conductive medium—creates all sorts of problems. After all, there's a form of wireless electrical transmission that occurs naturally. It's called lightning. Lightning isn't that useful to us, though, what with the way it kills people and sets fire to things. And occasionally blows up our other electricity infrastructure.

But that's not to say the concept is impossible. Scientists have known for many years that there's a massive amount of electricity just floating around freely in the atmosphere. The Earth itself has a negative electrical charge, and the atmosphere has an increasingly positive charge the higher up you go. All we need to do is figure out a way to direct power to where we want it.

There are a number of proposals, some involving microwaves, that seem pretty terrifying. But one man, as long ago as 1899, came up with what he thought was a safe and effective wireless transmission system.

His name was Nikola Tesla, and he invented such things as a practical generator to produce the "alternating current" we use in our homes today. He proposed and demonstrated a wireless transmission system that he claimed could have provided the world with free electricity nearly 100 years ago.

He went bankrupt. Conspiracy from big business? Was the science wrong? We don't know—but experiments are ongoing.

Electromagnetic Induction

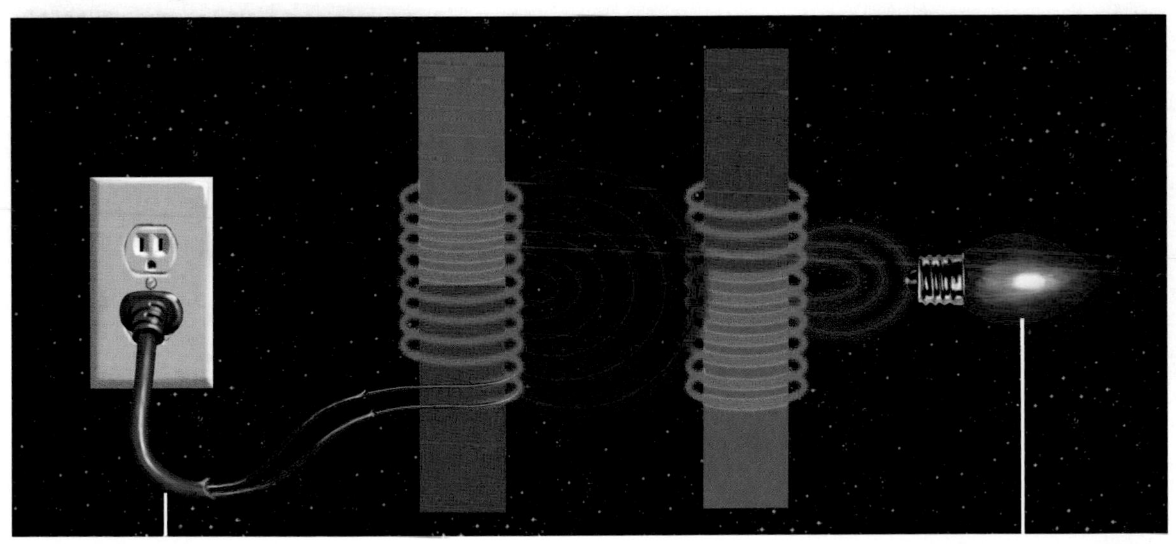

Q:

Why can't I survive a 200-foot fall into water?

Jumping into water from a few feet up is fun, but jumping off a large bridge into water is deadly. Why does the speed of impact make so much difference?

A:

The faster you hit the water, the less time it has to get out of your way, and the more like hitting a solid surface it becomes. It's thanks to a property called "cohesion" ....

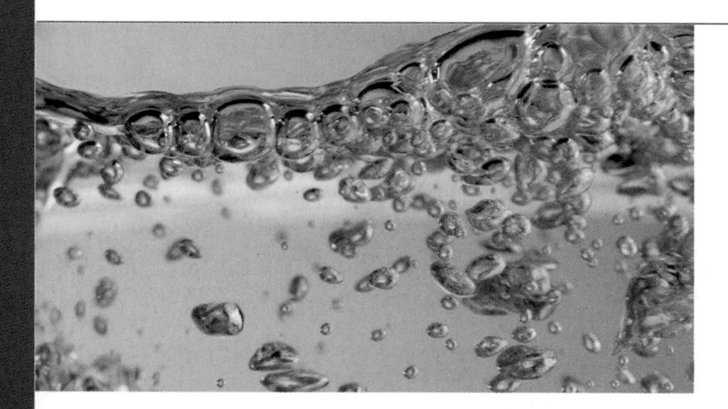

Don't believe the movies. If you fall into water from a height much over 80 feet, you probably won't survive. At the very least you'll be terribly injured, almost as if you'd dropped onto solid cement.

Before we get to the water, let's look at why smashing into something at high speed can be fatal. According to the basic physical laws of the universe, if a body is moving, it has been charged with a particular kind of energy, called kinetic energy.

To slow the body down or even stop it, that kinetic energy has to be transferred into another body. The safest way to stop your bike is to apply the brakes—friction takes your kinetic energy and turns it into heat, and transfers some of it through the wheels of the bike into the ground. Eventually all your kinetic energy is gone, and you've come to a safe stop.

Hitting a brick wall also stops you, but your kinetic energy tries to transfer all at once into the bricks. The bricks won't accept very much of this energy at all, so it gets transferred back into your body. Basically, your forward motion bounces off the bricks and surges back through your body. Unfortunately for you, this surge is strong and chaotic enough to rupture blood vessels, destroy tissue, and even break bones.

It could even cause your brain to bounce back and forth inside your skull as all that kinetic energy is dissipated. This can lead to a fatal brain injury. Or you could have a heart attack, or rupture an artery and bleed to death.

Water is a good substance to hit at low speeds because it's a liquid. The molecules in the liquid are not locked into crystal lattice like in a solid and can flow out of your way. When you push your hand into a bucket of water, the molecules are pressed to the side and the level of the water in the bucket rises in proportion to how much weight you are using to press down.

But water has a special property called "cohesion." This means the water molecules prefer to stick together—which is why water forms beads and drops, and why small insects can actually skate on the water's surface. This cohesion is also called "surface tension."

One of the effects of surface tension is that water needs time to move out of your way. It can't do it instantly. The harder you hit the water, the more it will resist your body pushing through the surface.

A painful belly-flop at the pool is a harsh reminder of why jumping off tall objects into water is a really bad idea. Hit water hard enough and kinetic energy will be bounced back up through your body. If you're lucky, you'll end up with a broken ankle. If you're unlucky, you'll be killed instantly.

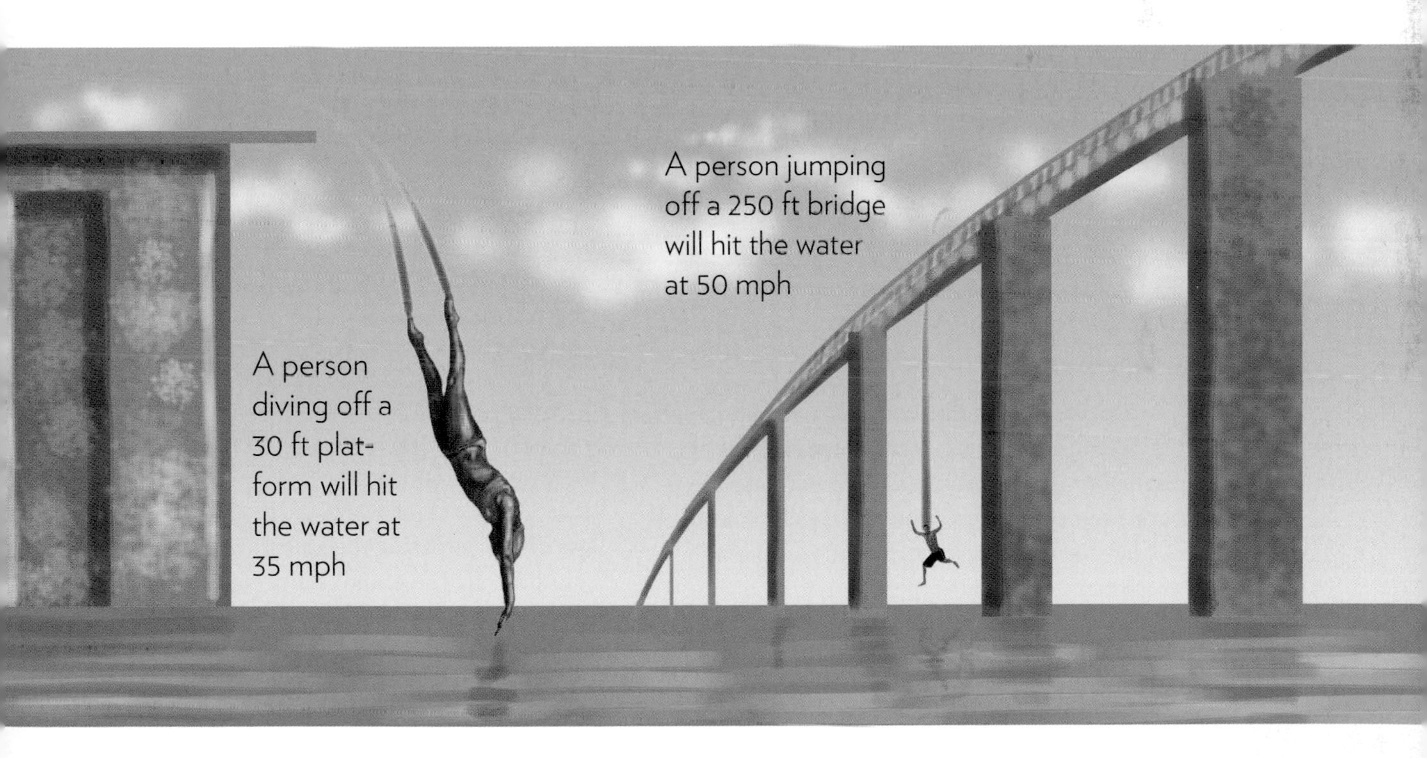

Q:

Why is a metal spoon colder than a plastic spoon?

Some materials like marble and most metals feel cool no matter how hot a day it is. How do these things stay so cool?

A:

Metal and stone aren't necessarily cool, they just *feel* cool because they're good at sucking heat away from your fingers.

Oddly enough, your skin doesn't really detect what temperature an object is. Rather, it detects how much heat is flowing from that object into your skin—or indeed how much heat the object is sucking out of you.

What's the difference between that and actual temperature? An object's temperature has to do with how much energy is stored inside it. A stone or a piece of wood could be very hot, but if that heat doesn't flow into our skin, the cells inside can't detect it.

One of the fundamental laws of physics—the Second Law of Thermodynamics—says that if you have two objects and one is hotter than the other, the heat will flow into the colder object until both objects are the same temperature.

But some materials let heat flow through them more easily than others. If you wrap an ice pack in thick Styrofoam, the heat from the air can't get to the ice as easily, and it takes a lot longer to melt. If you wrap your hand in a mitten, you can make snowballs without getting frostbite, because the mitten stops heat flowing from your hand into the snow.

This property is called "conductivity," and most metals are like the opposite of mittens—they have very good conductivity. When you touch a metal like iron, the heat from your finger flows into the metal and is quickly drawn away. Because of this, the sensors in your skin detect a lack of heat, and so give you a signal saying the metal is cold. Or at least very cool.

But the temperature of the metal might be more or less the same as the air in the room. It's just because the heat from your hand gets drawn away so quickly that your brain thinks it's touching something that really is cold.

A plastic spoon, on the other hand, does not conduct heat as well as a metal one, so it won't feel as cold to the touch.

The reverse of these actions is true as well. If you put a metal spoon into hot soup, it will "suck up" a lot of heat from the soup and become very hot—maybe even hot enough to burn your mouth. But a plastic spoon in the same soup probably won't get too hot to suck on, because not as much heat will have flowed into the plastic.

This is why we usually cook in metal pots and pans—because heat from a burner or electric element flows into the metal and gets distributed very evenly through the base and sides of the pot.

So how do we really know what temperature an object is? Well, without special equipment we don't, but then there's usually no reason we need to know. Our temperature sensors are designed to warn us when too much heat is flowing into or out of our skin. Since too much heat coming in can burn, and too much flowing out can freeze, that's the most important information for our sense of touch to communicate.

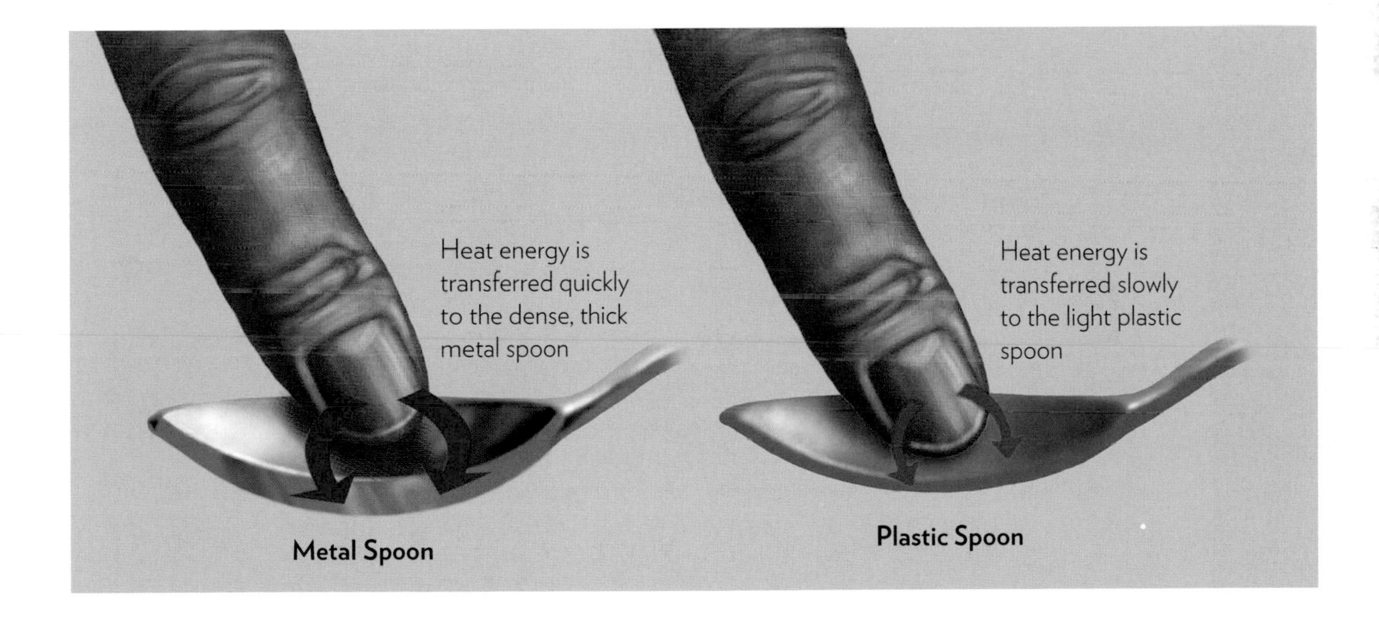

Q:

Why do tsunamis only become so destructive close to land?

When a big earthquake sets off a tsunami, we hear of boats out at sea just bobbing up and down slightly, while the shoreline gets totally destroyed. How is this possible?

A:

When a tsunami hits land, it "bunches up" and all its energy piles onto the coast at once, with catastrophic results ....

Waves rolling softly on the beach—is there any sound more soothing or relaxing? A wave is really just a swell of energy being transmitted through the water, passing from molecule to molecule.

This energy forms an undulating shape with a peak where the energy is highest and a trough where the energy is lowest. The distance between the peak and the trough is called the "wavelength" and the height of the wave above normal flat sea level is called its "amplitude."

When a wave moves across the ocean, it has to be able to transfer energy from one part of the water to the next. Out in deep water, it can do this quite gently and gradually. Closer to land, the water bumps up against the sandy bottom and against the beach.

Water behind the wave starts to bank up, making the wave shorter and taller. Eventually the wave gets so tall it collapses forward and breaks.

When this happens at the beach on a sunny day, it's a fun time for everyone. When it happens in a tsunami, it's a disaster.

A tsunami is an unusually large wave with lots of energy and power. Most waves are created by the wind blowing on the surface of the ocean. But a tsunami most often comes from an underwater earthquake, though landslides and volcanoes on the ocean floor can also trigger them.

Energy radiates outward from the sea floor and heads toward land. Way out at sea, the tsunami may be only a couple of feet high, but there's an immense amount of energy distributed from the surface to the seabed thousands of feet below.

As the tsunami gets closer to shore, there's less water to distribute the energy through, and so the wave starts to build up. The shallower the ocean, the higher the tsunami will grow.

Compared to waves generated by, say, a hurricane whipping up water with strong winds, tsunamis aren't usually that high. A typical tsunami is less like a gigantic surf wave crashing over a city, and more like a super-fast flash flood.

It's not just the height of the wave, but the sheer amount of water that surges inland that causes so much destruction—in other words, the wavelength. The energy stored in the tsunami is enough to push water miles across ground, smashing and uprooting anything not built specifically to withstand it.

After the energy of the wave is dissipated, the water is then pulled back into the ocean by gravity. After all, the only reason the sea stays where it is is because our islands and continents are much higher than the sea floor.

The wave can do more damage on the way out, and what's worse, gravity gives the water more energy. A tsunami can bounce around inside a harbor for a long time. The exact topography of the sea floor and surrounding coast can even split a tsunami into multiple waves that smash into the land one after the other.

Uncontrolled energy is a dangerous thing, especially when it crashes across a coastal city.

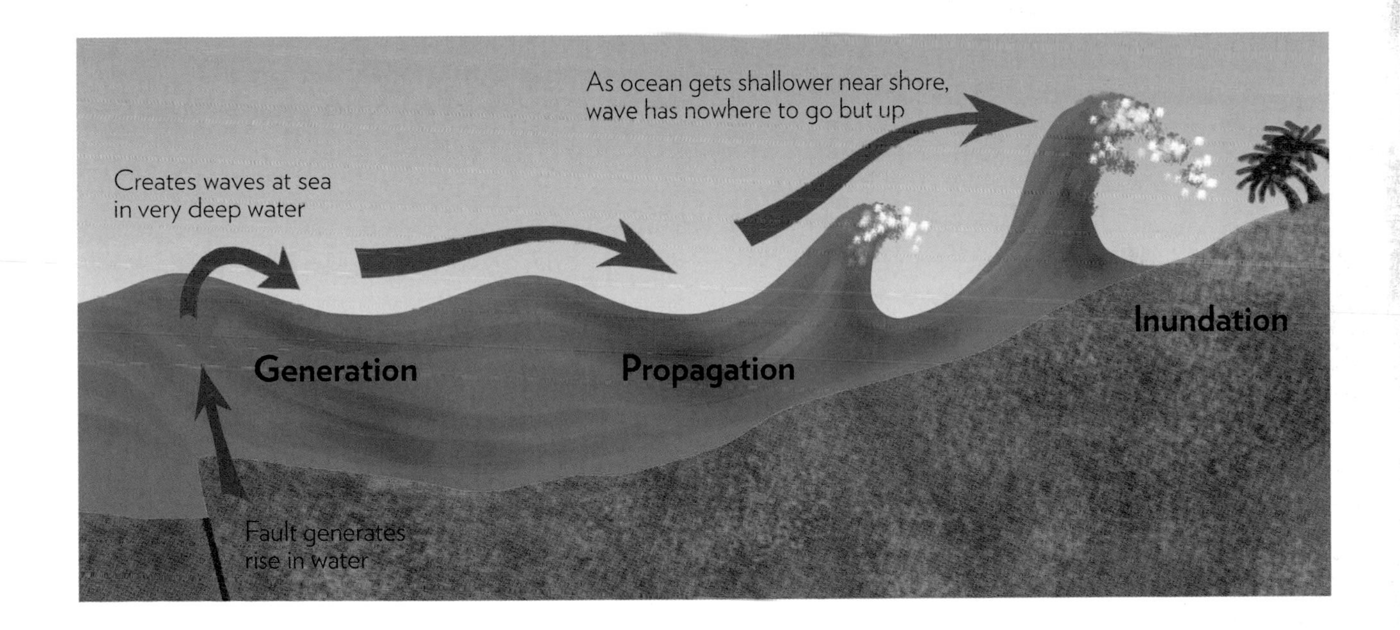

Q:

Why do I see the lightning flash long before I hear the thunderclap?

Thunder is the sound of the shockwave from a lightning strike, but thunder and lightning are always "out of sync." Why does it take so long to hear the thunder?

A:

Sound travels through air much slower than light. But you can hear the thunder from beyond the horizon ....

There's nothing like sitting on a porch and watching a thunderstorm play across a mountain range several miles away. From a distance, lightning flickers constantly and thunder is a continuous rumble.

Lightning causes thunder by ripping through the air and creating a shockwave. This shockwave spreads out in every direction and is eventually picked up by our ears. Close up, thunder "claps" or cracks and can be so loud it breaks glass or causes temporary deafness. Farther away, thunder sounds more like a rumble as the air absorbs energy from the shockwave and dampens it.

Farmers and outdoorsy folks know how to figure out the distance of a thunderstorm by counting the seconds between a lightning flash and a thunderclap.

They can do this because sound travels much slower through air than light does. Since the speed of light is so fast, you'll see the lightning virtually at the exact moment it strikes. How long it takes you to hear the thunder depends on the distance you are from the storm.

It's quite easy to calculate the distance, because sound takes about five seconds to travel 1 mile (1.6km). So you watch for the lightning and then start counting off seconds using a watch or cellphone. When you finally hear the thunder, divide the number of seconds by five. That's how far away the thunderstorm is in miles.

The explanation for this gap is straightforward: light travels through our atmosphere at nearly 100 percent of light speed—671 million miles per hour (1,079,869,824km/h). Sound travels at only 768 mph (1,236km/h).

That's because sound is what's called a "compression wave." It forms when an air molecule gets pushed by something (like air getting ripped apart by lightning), then it knocks against the next air molecule, which knocks against the next air molecule, and so on until the wave reaches your eardrum and your nerves pick up the change in air pressure.

Light, on the other hand, is made up of photons. These tiny subatomic particles are emitted, in this case by the lightning, and they travel through the air almost uninterrupted until they hit your retina.

While the speed of light is only a tiny bit slower in air than it is in the vacuum of space, sound changes speed very dramatically, depending on what it's passing through.

The speed of sound in seawater, for instance, is a whopping 3,490 mph (5,616km/h). This is important for animals like whales, which communicate over vast distances by singing. Perhaps it's no coincidence that one of the loudest creatures on the planet, the sperm whale, lives in the ocean. A sperm whale can generate a pulse of sound as loud as 230 decibels underwater (equivalent to 170 decibels in air), which is louder than a jet engine or someone firing a gun right next to your ear.

Because of the way sound moves through the air, you don't need direct line of sight to hear something—it just depends on the sensitivity of your hearing. Many animals can hear the low rumble of approaching thunder long before the storm comes over the horizon. And young children are often much better at detecting very distant sounds than older people.

So your dog doesn't really have a "sixth sense" about thunderstorms. He hides under the bed because he can already hear the thunder!

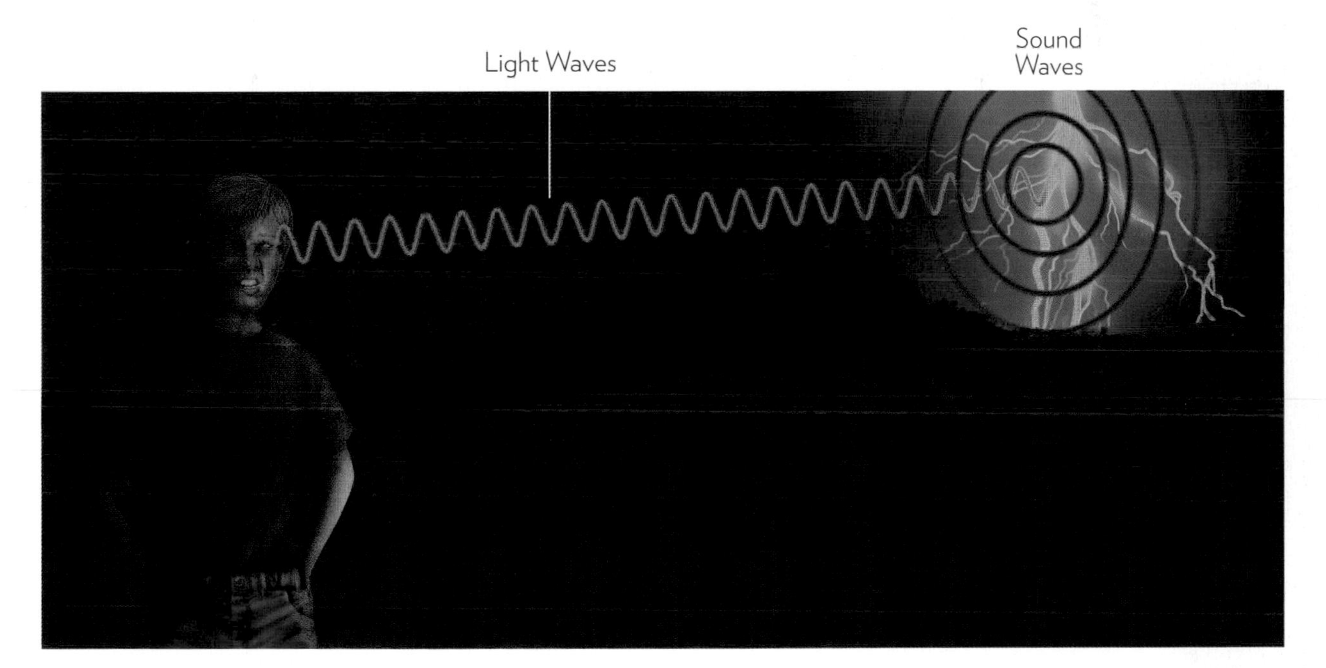

In the time it takes the light waves to reach the person, the sound waves travel only a fraction of the distance

Q:

Why do I float more easily in the ocean?

Staying afloat in the sea is less tiring than trying to float in a lake or river. How is this possible?

A:

You float if you can displace the amount of water that equals your weight before your head goes under. And seawater weighs more than freshwater, so you would need to displace less ....

Swimming can be a tricky business. If you stretch out your arms and legs and lie on your back, you should be able to float. But if you point your feet straight down into the water, you'll sink.

It's all to do with your density—how much you weigh relative to how big you are. When we go swimming, we're being pulled down into the water by gravity. The size of the area right underneath us changes, depending on whether we're pointing straight down (small area) or lying horizontal (large area).

As far as physics and water are concerned, if you present a large surface area, you're effectively less dense than if you dangle your legs into the water or curl up into a ball. You can gain a little extra buoyancy by taking a deep breath and filling your lungs with air—this reduces your average density, but obviously you won't be able to keep this up for long.

How much you sink depends on what you weigh and how big you are. You will sink into the water until you've displaced—or moved aside—an amount of water equal to your weight. If you are less dense than the water, then the volume or size of the water displaced will be smaller than you—and part of your body will then stick out of the water. We suggest the head.

The denser the water, the more you will float. Cold water is denser than warm, so it's easier to float in icy lakes than in a tropical river.

Seawater is quite a bit denser than freshwater because it has salt and other minerals dissolved in it. Each liter of seawater has 35 grams of salt in it, and these salts don't really make a difference to the volume of the water. So the water is about 2.5 percent heavier than freshwater.

If you get a box 1 cubic foot (30.5cm) in size, it needs to weigh less than 62.4 pounds (28kg) to float in freshwater. But to float in saltwater, it can weigh as much as 64 pounds (29kg)!

A fun experiment would be to get such a cubic-foot box and weight it to exactly 63 pounds (28kg). The box would float (barely) in seawater, but sink in freshwater.

Salinity varies slightly throughout the world's ocean, but some isolated bodies of water have very high salinity. The Dead Sea is a lake on the border between Israel and Jordan and is famous for having extreme salinity. Nearly 10 times saltier than the ocean! Humans have known about and used the lake's unusual chemistry for thousands of years, for everything from salt production to the world's first health resorts.

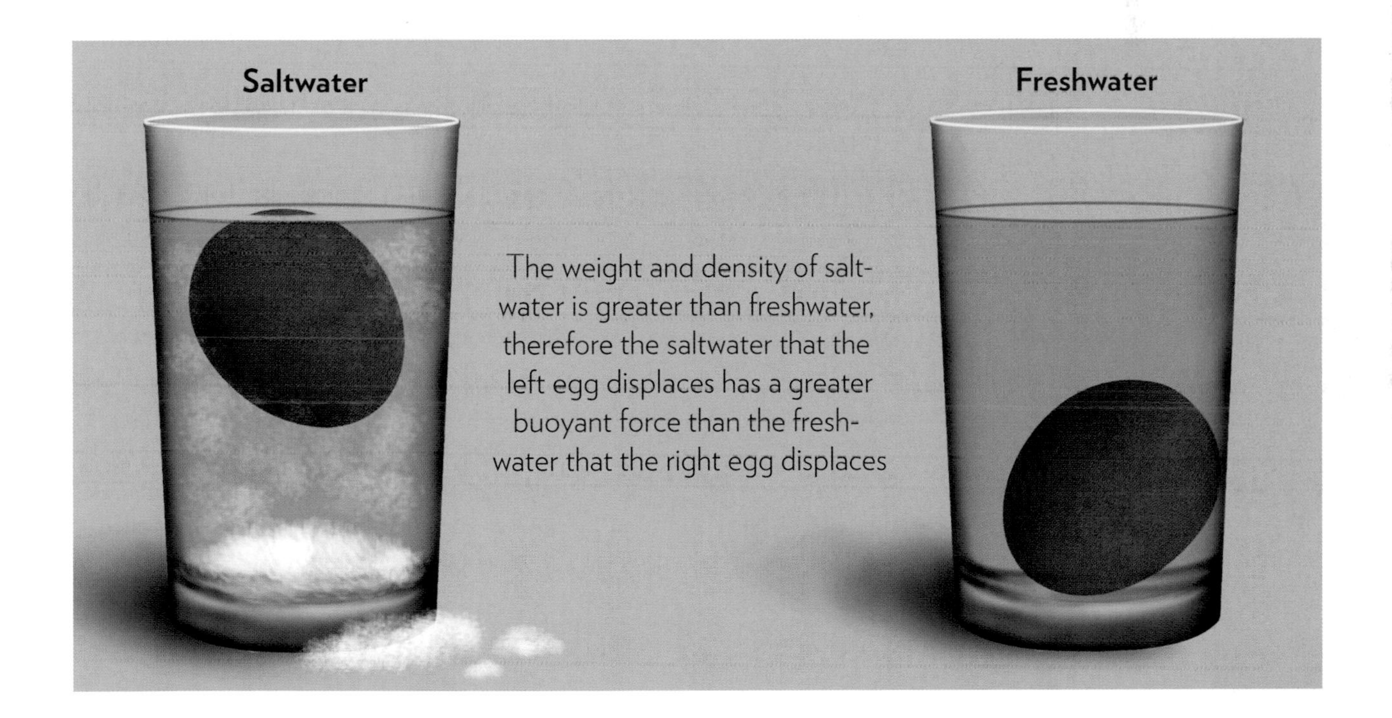

The weight and density of saltwater is greater than freshwater, therefore the saltwater that the left egg displaces has a greater buoyant force than the freshwater that the right egg displaces

Q:

Everyone knows hot air rises ... but why does it?

The way hot air rises and cold air sinks is what makes hot air balloons and space heaters possible—and it also drives our weather systems. But why does hot air rise?

A:

For the same reason that a boat rises through water to float—because it's less dense.

Storms develop in our atmosphere because hot air rises and cold air rushes in to fill the space left behind. This can draw water up from the oceans, create winds, and eventually produce rain. Thank goodness this system works, because it keeps the water cycle turning and makes life on land possible!

Obviously, our atmosphere is a gas—or more accurately, a mix of a few different gases. A gas is a bunch of molecules that are not stuck together, and so bounce around chaotically. If you put them in an empty container (and we mean really empty—as in a vacuum), the gas molecules will expand to fill the whole of the container.

When you mix gases, things get a bit more complicated. Some gases have bigger molecules than others, so they are denser. Gravity will pull a denser gas to the bottom of a container, while a less dense gas will accumulate at the top.

Another way to change the density or pressure of a gas is to heat it. When heat enters a gas, the molecules become more energetic and bounce around more. The gas expands and squishes the colder denser gas farther toward the bottom of the container.

Our atmosphere is in a container made of gravity, with an open top leading out into deep space. The farther up you go, the less dense the atmosphere becomes until it eventually just peters out.

Gravity keeps the atmosphere stuck to the surface of the planet, but less dense gases do have enough energy to rise higher into the sky than denser gases.

One way to explain why this happens is to think of the atmosphere as something you can float in. It's easy to do: just fill up a balloon with helium until your apparent weight is less than 1.2 pounds (.5kg) per cubic foot. Then you can just float up into the sky.

This is what happens to hot air. As it gets energy, it expands and takes up more volume. This changes its average density. When its density becomes less than that of the surrounding cold air, it becomes buoyant and will float upward.

The way hot liquids and hot gases float or rise is incredibly important. That's because the exchange of heat from the ocean to the atmosphere and back again is what drives the engine of the Earth.

Without hot air rising, there would be no weather. And without weather, there would be no water cycle. And without a water cycle taking water from the oceans, removing the salt, and dropping it over land, there could be no life outside the sea.

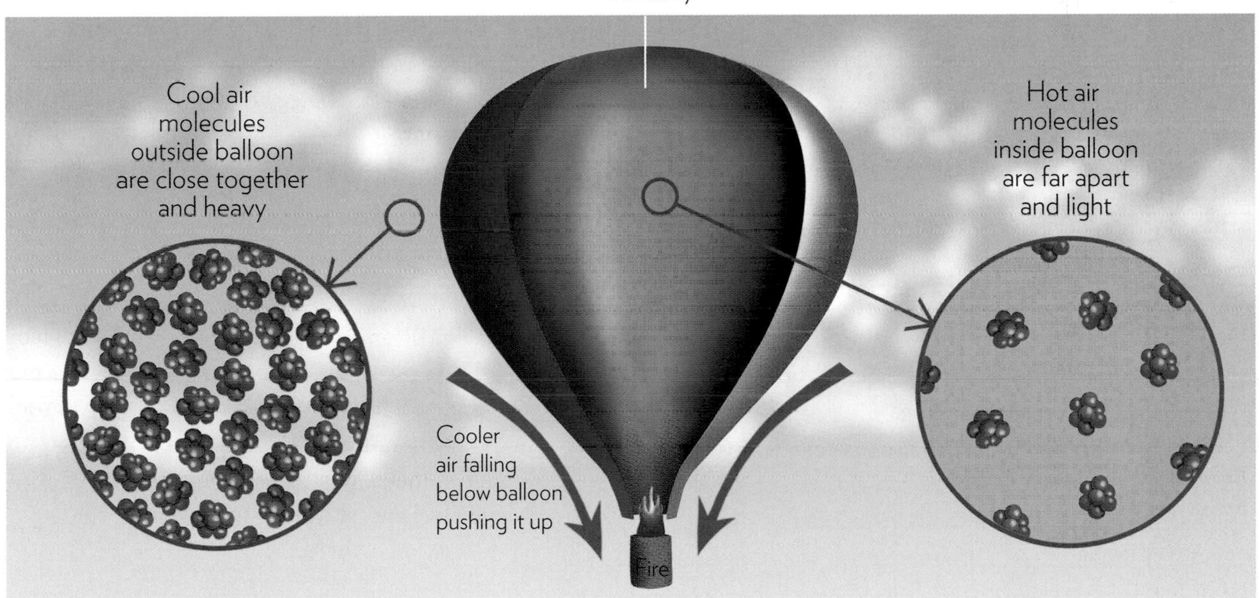

Q:

How does gravity work?

Gravity pulls us toward the center of the Earth, and the farther we go into space, the weaker Earth's gravity becomes. But how does all this actually work?

A:

Nobody knows for sure. We understand the laws of gravity well enough, but fitting it into our explanation of the rest of the universe is proving very difficult.

Gravity is, as far as we know, one of the fundamental forces of nature. If two objects are made of normal matter and they have mass (what we call weight here on Earth), then they are attracted to each other by gravity.

The strength of this gravitational pull depends on how much mass the objects have and how far apart they are. The closer and heavier the greater the gravitational attraction.

When one object is very massive and the other is very light—such as the Earth versus your body—then gravity feels like the smaller object is sticking to the larger. You stick to the Earth, or rather the Earth constantly tries to pull you down into the core. Only the solid crust gets in the way.

And this brings up something very odd about gravity. It's a fundamental force that acts on everything in the universe, but it's actually really weak compared to the other forces.

The gravitational pull of a planet weighing many trillions of tons is not strong enough to pull you through a layer of rock that we can crack quite easily with a few sticks of dynamite. Earth's gravity isn't strong enough to stop your puny human muscles from being able to resist it and raise your arm, or throw a ball into the air.

The other fundamental forces—the strong nuclear force, the weak nuclear force, and electromagnetism—are all of roughly equal strength. What's more, when you start messing around with quantum mechanics, it turns out these three forces are all different aspects of the same thing. They can be unified.

But gravity can't fit into this system. It stands apart, obvious but inexplicable.

Is there a particle especially for gravity, like a photon is for light? No one knows, yet. There are many theories, including Einstein's, which says that gravity isn't really a force, but evidence of the way space and time curve the closer they are to very massive objects like planets and stars.

Experiments in measuring the Earth's gravitational field are—compared to quantum mechanics, anyway—very straightforward. We have a detailed map of the way gravity fluctuates across the surface of the planet according to how dense the rock is underground. It's true: you weigh about 0.7 percent more in Helsinki than you do in Singapore due to the variation in density.

Gravity has also led to the discovery of one of the biggest mysteries in science—dark matter. When we apply our understanding of gravity to our observations of how galaxies rotate, it seems there is not nearly enough matter or mass. Further observations and experiments suggest that as much as 84.5 percent of the "stuff" in the universe is made of dark matter. We can't see it or interact with it, but figuring out its true nature is a major focus for physicists and cosmologists working today.

Gravity can be described as a curve in space caused by the presence of a massive object (like Earth). Objects travel along the curve, which makes it seem like they are being pulled toward the center of Earth's gravity

This is Einstein's explanation of how gravity works. But physicists continue to explore other possibilities, including that gravity might have its own particle called a graviton.

index

Numbers

A

B

C

D

E

F

G

H

I

J–K

L

M

N

O

P

U

V

W-X-Y-Z

Photo Credits:

The topographical map of sea level change on page 51 courtesy of:

GLOBE Task Team and others (Hastings, David A., Paula K. Dunbar, Gerald M. Elphingstone, Mark Bootz, Hiroshi Murakami, Hiroshi Maruyama, Hiroshi Masaharu, Peter Holland, John Payne, Nevin A. Bryant, Thomas L. Logan, J.-P. Muller, Gunter Schreier, and John S. MacDonald), eds., 1999. The Global Land One-kilometer Base Elevation (GLOBE) Digital Elevation Model, Version 1.0. National Oceanic and Atmospheric Administration, National Geophysical Data Center, 325 Broadway, Boulder, Colorado 80305-3328, U.S.A. Digital data base on the World Wide Web (URL: http://www.ngdc.noaa.gov/mgg/topo/globe.html) and CD-ROMs.